COMBINE HARVESTING

FUNDAMENTALS OF MACHINE OPERATION
FMO15104NC

How to operate, maintain, and improve
the efficiency of your combine.

FREE CATALOG
Call 1-800-522-7448

Check Out <u>All</u> Of Our Titles In The FUNDAMENTALS OF MACHINE OPERATION Series!

Visit Us on the Internet--
http://www.deere.com/aboutus/pub/jdpub/

Here are a few of our titles in this series:

Tractors
How to operate, maintain, and improve the performance of your tractor

Tillage
A practical guide to the latest tillage methods, conservation planning, crop residue management, and solution to soil compaction

Hay and Forage Harvesting
"Real world" evaluations of the many different ways to improve hay and forage harvesting and storage efficiency

Preventive Maintenance
A practical step-by-step guide to reducing equipment operating costs and downtime and improving operating safety

Planting
A practical step-by-step guide to increasing planter, seeder, and drill machine capacities; improving their field efficiencies; and maximizing yields

Combine Harvesting
How to operate, maintain, and improve the efficiency of your combine

Machinery Maintenance
A practical guide using basic language to explain good machinery maintenance to an international audience

Our FUNDAMENTALS OF MACHINE OPERATION Series
shows you much more than how to safely operate equipment. You learn machine capacities and adjustments, field efficiency, how to improve machine performance, and how to eliminate unnecessary field operations.

Call 1-800-522-7448 to order; to inquire into prices; or to get our free catalog.

Check Out <u>All</u> Of Our Titles In The FUNDAMENTALS OF MACHINE OPERATION Series!

Tractors
How to operate, maintain, and improve the performance of your tractor

Tillage
A practical guide to the latest tillage methods, conservation planning, crop residue management, and solution to soil compaction

Hay and Forage Harvesting
"Real world" evaluations of the many different ways to improve hay and forage harvesting and storage efficiency

Preventive Maintenance
A practical step-by-step guide to reducing equipment operating costs and downtime and improving operating safety

Planting
A practical step-by-step guide to increasing planter, seeder, and drill machine capacities; improving their field efficiencies; and maximizing yields

Combine Harvesting
How to operate, maintain, and improve the efficiency of your combine

Machinery Maintenance
A practical guide using basic language to explain good machinery maintenance to an international audience

Our FUNDAMENTALS OF MACHINE OPERATION Series shows you much more than how to safely operate equipment. You learn machine capacities and adjustments, field efficiency, how to improve machine performance, and how to eliminate unnecessary field operations.

Call 1-800-522-7448 to order; to inquire into prices; or to get our free catalog.

Combine Harvesting
Operating, Maintaining and Improving Efficiency of Combines

Fundamentals of Machine Operation

PUBLISHER

Fundamentals of Machine Operation (FMO) is a series of manuals created by Deere & Company. Each book in the series was conceived, researched, outlined, edited, and published by Deere & Company. Authors were selected to provide a basic technical manuscript which could be edited and written by staff editors.

PUBLISHER: DEERE & COMPANY SERVICE PUBLICATIONS, Dept. FOS/FMO, John Deere Road, Moline, Illinois 61265-8098; Dept. Manager: Alton E. Miller.

SERVICE PUBLICATIONS EDITORIAL STAFF
Managing Editor: Louis R. Hathaway
Editor: John E. Kuhar
Promotions: Lori J. Lees

AUTHOR: *George A. Griffin* is a retired supervisor of the Service Publications Department at John Deere Harvester Works. He has had over 25 years of experience with combines. The majority of this experience has been in writing and teaching the operation, maintenance and service of John Deere Combines.

CONTRIBUTING AUTHOR: *Frank Buckingham* is an agricultural engineer and freelance writer of numerous articles on agricultural machinery.

CONTRIBUTING AUTHOR: *Thomas J. Meyer* has an accumulated 25 years of technical writing experience. Mr. Meyer has authored numerous publications on both agricultural and industrial machinery. He has been instrumental in the training and development of several aspiring technical writers during his career.

CONSULTING EDITOR: *Thomas A. Hoerner*, Ph.D., Associate Professor and Teacher Educator in Agricultural Mechanics at Iowa State University. Dr. Hoerner has 24 years of high school and university teaching experience. He has authored numerous manuals and instructional materials in the Agricultural Mechanics field.

CONSULTING EDITOR: *Ralph J. Moens* is a teacher of vocational agriculture at the high school in Atkinson, Illinois. A part-time farmer, Mr. Moens has 26 years of experience in preparing young people for agribusiness careers.

CONSULTING EDITOR: *Keith R. Carlson* has thirteen years of experience as a high school vocational agriculture instructor. Mr. Carlson is the author of numerous instructor's guides, including the American Oil Company's "Vo-Ag Management Kits." Many of his aids have been available under the name of Vo-Ag Visuals. All instructor's kits for the FMO texts are being prepared under the direction of Mr. Carlson, who is presently General Manager of Agri-Education, Inc.

CONSULTING EDITOR: *John A. Conrads,* a retired John Deere service executive has had over 30 years of hands-on experience with farm and industrial machinery. He has authored and edited numerous articles on the subjects of service, training and effective technical communication. He is also a well known speaker and seminar leader.

SPECIAL ACKNOWLEDGEMENTS: The author wishes to thank the following John Deere people for their assistance in preparing this book: Mr. D. C. Bichel, Mr. R. M. Poterack, Mr. J. L. Crow, and Mr. M. H. Krouse.

CONTRIBUTORS: The following persons and groups were very helpful in giving valuable literature and technical assistance: Duetz-Allis Corporation, Batavia, IL; American Association for Vocational Instructional Materials; American Society of Agricultural Engineers; Case-IH, Chicago, IL; Iowa State University; Ohio State University; Ford-New Holland, New Holland, PA; United States Department of Agriculture; and University of Illinois. We also wish to thank a host of John Deere people who gave extra assistance and advice on this project.

FOR MORE INFORMATION: This text is part of a complete series of texts and visuals on agricultural machinery entitled Fundamentals of Machine Operation (FMO). For information, request a free FMO Catalog of manuals and visuals. Send your request to John Deere Service Publications, Dept. FOS/FMO, John Deere Road, Moline, Illinois 61265-8098.

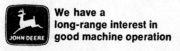

We have a long-range interest in good machine operation

Copyright © 1973, 1981, 1987, 1991/Fourth Edition/Deere & Company/Moline, Illinois/All rights reserved

ISBN-0-86691-132-4

Contents

1
INTRODUCTION

What is a "Combine?" 2
Major Developments Leading to Modern Combine 3
Combine Types and Sizes 4
Self-Propelled Combines 4
Level-Land Combines 11
Hillside and Sidehill Combines 11
Combine Sizes 12
Pull-Type Combines 14
Windrowing Method 15

2
HARVESTING SYSTEMS

Introduction 25
Cutting and Feeding the Crop 25
Reel Fore/Aft Operation 28
Threshing the Crop 34
Types of Threshing Systems 35
Threshing Cylinder, Rotor, and Concave Function 38
Threshing Cylinder, Rotor, and Concave Adjustments 39
Effects of Separator Speed and Cylinder- or
Rotor-Concave Settings 41
Determining Threshing Action 42
Separating the Crop 43
Cleaning the Crop 49
Handling the Crop 55

iii

3
POWER SYSTEMS

Introduction	60
Engines	61
Engine Types	62
Operation of Engines	63
Engine Systems	65
Transmitting Engine Power	71
Power Trains	77
Hydrostatic Drives	82
Four Wheel Drive	87
Leveling System (Hillside Combine)	89

4
OPERATING CONTROLS

Introduction	94
Identifying Controls	94
Controlling Combine Movement	97
Starting the Engine	99
Starting Tips	100
Driving the Combine	100
Transporting the Combine	105
Controlling Combine Field Operation	106
Electrical Accessories	112
Symbols for Operating Controls	116

5
FIELD OPERATION AND ADJUSTMENTS

Introduction	120
Proper Operation and Adjustments	121
Factors Affecting Combine Harvesting	122
Planning and Preparation	123
Harvesting Methods	123
Grain Windrowers	124
When to Combine	132
Production Capacities of Combines	133
Mechanical Condition of Combine	134
Preliminary Settings of Combine	137
Field Operation and Adjustments	142
Operating and Adjusting Threshing Units	151
Determining Grain Losses	154
Field Conditions	164
Troubleshooting Field Problems	166

6

MAINTENANCE AND SAFETY

General Maintenance	182
Engine and Power Train Maintenance	183
Header Maintenance	185
Threshing Unit Maintenance	189
Separating Unit Maintenance	191
Cleaning Unit Maintenance	192
Grain Handling Maintenance	192
Wheel and Track Maintenance	193
Belt and Chain Maintenance	195
Combine Storage	197
Safety	200

SUGGESTED READINGS 210

APPENDIX 211

INDEX 213

1
Introduction

Fig. 1—The Harvester-Thresher Of Grandfather's Day (20 Mule Team)

WHAT IS A "COMBINE?"

Early combines were known as "harvester-threshers" — machines that were pulled through the fields by teams of horses or mules as shown in Fig. 1. Some machines were known only as "threshers" (Fig. 2) because the grain was first cut and then brought to a thresher for threshing and separating the grain from the straw.

Later the thresher was powered by a steam engine or tractor through a flat belt drive. Some harvester-threshers were propelled through fields by steam engines. These machines require a crew of men to harvest and thresh what one man can handle with a modern combine.

The combine, as we know it today, is a machine used to harvest and thresh all kinds of grain in a variety of crop and field conditions. The name "combine" developed when the harvesting and threshing operations were "combined" into one complete machine.

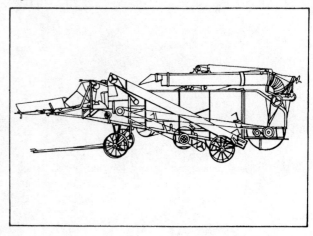

Fig. 2—The Thresher

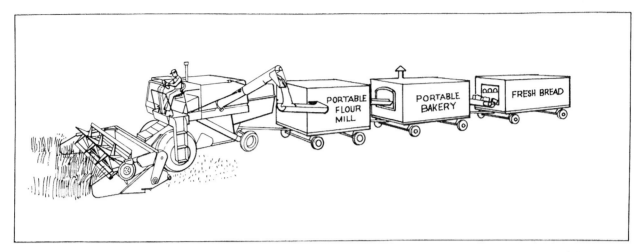

Fig. 3—The Ultimate In Combining Wheat

Shown in Fig. 3, is the ultimate in combining wheat — from field to completed product which is ready for the table. While this seems farfetched to us today, just a little more than a hundred years ago the concept of combining all harvesting and threshing operations into one machine seemed unbelievable.

Although many people may know what a combine is, few know the details of what it does or how it works. The combine and its operation are described in this book—but first let's look at the evolution of the combine from its humble beginning.

The following chart depicts some of the major developments leading to the modern combine. Many machines were invented during the periods shown; those listed here are only typical developments. Each milestone is indicated by a period in time when important progress was made. The years shown do not always indicate when the machine was invented, but rather tell when the machine made its greatest impact in agriculture. The column on the right tells what happened in history during the same period.

MAJOR DEVELOPMENTS LEADING TO MODERN COMBINES

Year	Mechanical Progress	What Else Was Happening Then
To 1800	*Hand tools* were the only means of harvesting grain for centuries. The sickle is the oldest tool and was used in our country during the Colonial Days. The scythe was developed later. It eliminated stooping and improved the speed of cutting. The cradle scythe was the first hand reaper which allowed the grain to be cut, collected and deposited in untied bundles.	The ancient Egyptians used sickles over 1400 years before Christ was born. The scythe was used during the Romans' rule of the world.
1800	Threshing was done by beating the grain from the straw with hand flails before the *"ground hog" thresher* was invented. This machine was a small stationary thresher that knocked the grain from the straw. The grain still had to be separated from the straw and chaff by hand winnowing.	Washington became the new capital of the United States.
1830's	The *McCormick reaper* mechanized the cutting and gathering of grain. It cut the grain and raked it from the platform into bunches. These bunches were collected onto a wagon by hand and taken to the "ground hog" for threshing. The Moore-Hascall Harvester was one of the first harvester-threshers built that performed the basic functions of cutting, threshing and cleaning grain.	Battle of the Alamo occurred. First steel plow invented by John Deere at Grand Detour, Illinois.
1864	The *Marvin Combined Harvester* was patented. It was a four-wheeled carriage type.	Civil War was being fought.

(Continued)

Year	Mechanical Progress	What Else Was Happening Then
1886	The *Hauser Harvester* was patented. The separator was driven by a geared ground wheel.	Statue of Liberty was unveiled.
1889	The *steam-powered combine* was patented by Daniel Best. The steam engine was mounted on the combine and a large steam tractor pulled the machine through the field.	The famous Oklahoma land rush occurred.
1892	Benjamin Holt patented a *hillside combine* which kept the separator on a level plane.	The first gasoline engine-powered automobile was made in the U.S. by Charles and Frank Duryea.
1900-1935	Binders were used extensively to cut and gather small grains.	Birth of mass automobile production. World War I fought.
1920's	*Tractor-drawn combines* were becoming common in the wheat belt.	Women's Suffrage (19th) Amendment to the Constitution was declared constitutional.
1930's	*Tractor-drawn combines* were becoming more popular across the continent because low-cost machines were developed.	The Great Depression occurred.
1940's	*Self-propelled combines* came into popular use.	World War II ended. The United Nations was formed.
1950-1970's	*Sophisticated self-propelled combines* were developed. Machines became gradually larger and more efficient. In 1955 self-propelled combines were adapted to harvesting corn.	America was experiencing the greatest technological advancements in history.
1975	First modern "rotary" combine marketed in the United States. John Deere introduced a sidehill combine.	Apollo/Soyuz rendezvous in space — a first for manned space craft built and launched by two different countries.

The harvester-threshers of yesterday were cumbersome machines that required very experienced operators. And they lacked the capacity, efficiency and crop flexibility that we have in our present machines.

The **modern combine** not only performs its basic functions better, but it also makes operation easier, safer, and more comfortable because:

- **Convenient controls** allow the operator to change speeds and settings from the operator's seat.
- **Hydraulic power** allows the operator to move heavy loads by simply moving a lever.
- **Shaft-speed-monitoring devices** permit him to check the operation of his machine at a glance.
- **Operator's cab** provides temperature control to keep the operator comfortable in any kind of weather and shields him from machine noise.
- **Refinements** have been made to make the combine as trouble-free as possible. Not only has operation been made easier, but maintenance has been reduced.

The development of the combine increased the harvesting of wheat, for example, from fractions of an acre per day with hand tools to 75 or more acres a day with modern combines. Let's take a look at some of the types and sizes.

COMBINE TYPES AND SIZES

Modern combines are available in a wide range of types and sizes. Technology has given us a machine that is very flexible — a combine that can harvest many types of crops under various field and crop conditions.

Combines can be categorized as either *self-propelled* or *pull-type* machines. The self-propelled combine can further be described as either a *level-land* combine or a *hillside* combine. Self-propelled combines are available in a variety of models, depending on the crop to be harvested. Pull-type combines are usually designed only for small grains, such as wheat or oats.

Combines may also be classified as **conventional**, with cylinder threshing and straw walker separation

Fig. 4—Some Of The Implements and Machines Leading To The Modern Combine

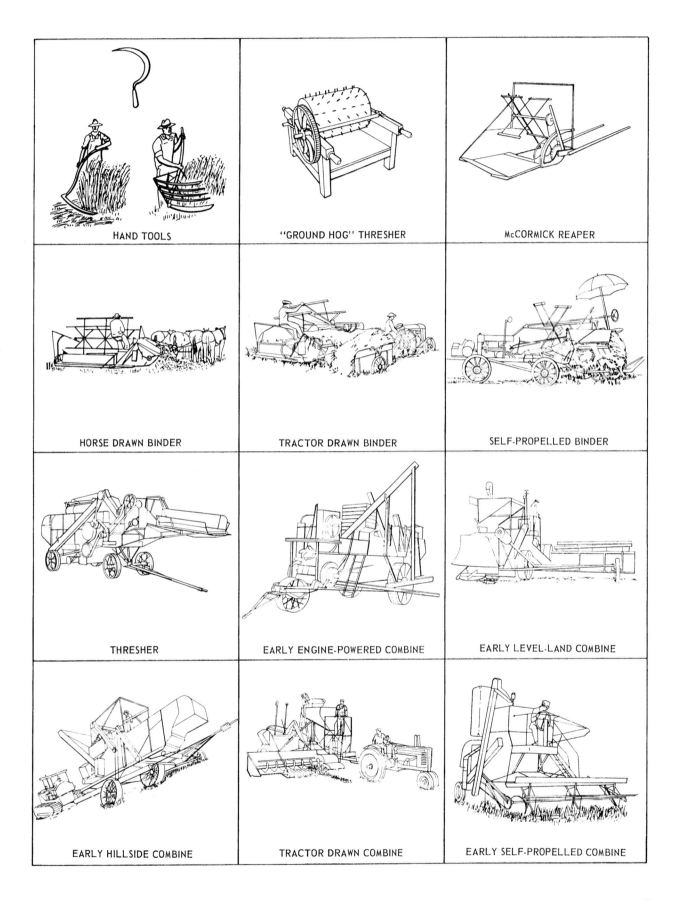

Fig. 5—Modern Combine

systems; or as **"rotary"** (sometimes called axial), in which a rotor or rotors replace the cylinder and straw walkers for threshing and separation of grain and straw. These differences will be explained in Harvesting Systems, Chapter 2.

SELF-PROPELLED COMBINES

The self-propelled combine didn't become widely available until the late 1940's. Many different types of self-propelled models were experimented with in those early years. Finding the right combination of engine horsepower and transmission of power was a difficult task.

Now this machine has been developed to where it has a husky engine that provides plenty of power to propel the machine through the toughest fields while combining the highest-yielding crops. It is also a machine operated by one man rather than a team.

Self-propelled units can harvest a swath up to 30 feet (9.1 m) wide in small grain or soybeans or up to 12 rows of corn.

The operator, seated high on the machine, has a clear, direct view of his work with all controls conveniently

Fig. 6—Controls Are Conveniently Located

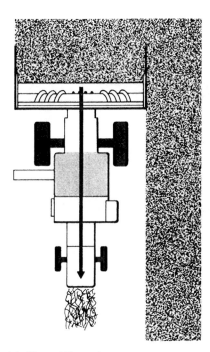

Fig. 7—Straight-Through Harvesting

located so that he can change the operation of the combine to adapt to the changing field and crop conditions. Most modern combines are equipped with cabs that protect the operator from cold, heat, dust and rain.

The fundamental design of the self-propelled combine is a *straight-through type* with the cut grain being delivered to the center of the platform and then up into the threshing unit (Fig. 7). This means that no standing grain is run down when opening a field. The operator may start combining anywhere, leaving green spots in the field to be harvested later, if desired.

The self-propelled combine is especially adapted for harvesting the large acreages of the Great Plains (Fig. 8). Special equipment is available to meet conditions as they exist in most territories and crops. It is a machine that is used in all parts of the world wherever there is sufficient acreage to warrant its use.

Specific models are available for harvesting such crops as rice, edible beans, corn, soybeans, and grass seed. These models are determined by the special type of equipment used on the combine.

For example, a rice combine will be equipped with different threshing parts than a corn combine. Also the rice combine will have special rice tires or it may even be equipped with crawler tracks. An edible-bean combine has special features for eliminating dirt from the beans. A corn and soybean combine uses one header to gather the corn and another header to gather soybeans. Sacker attachments are used when it is desired to put grain or grass seed into sacks instead of hauling it in bulk to the market or storage.

As mentioned, the basic types of self-propelled combines are:

- **Level-Land Combines**

- **Hillside Combines**

Level-land combines are used in flat areas or on slightly rolling hills. Hillside models are almost exclusively used on steep slopes such as those of the Pacific Northwest. Sizes vary considerably; this subject is discussed later in this chapter.

- **Sidehill Combines**

The sidehill combine (Fig. 10), introduced in 1975 by John Deere, is unique in that it levels to approximately 50 percent of what a hillside combine does. It was designed for use on slopes of up to 18 percent. This type of farm terrain is located primarily in the midwest and a few other areas.

Fig. 8—Harvesting On The Great Plains

Fig. 9—Handling Grain

Fig. 10—Sidehill Combine

Fig. 11—Combining Corn

Fig. 12—Combining Rice

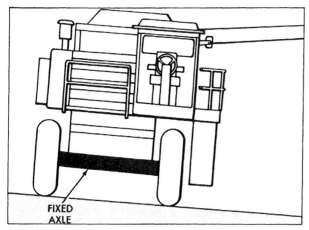

Fig. 13—Level-Land Combines Have a Fixed Drive Axle

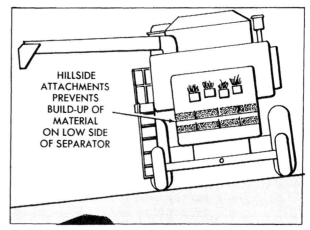

Fig. 15—Sidehill Attachments For Level-Land Combines Prevent Buildup Of Material On Low Side Of Separator

Level-Land Combines

The level-land combine is supported by a fixed drive axle (Fig. 11). When the combine operates over rolling ground, the separator tilts with the contour of the ground.

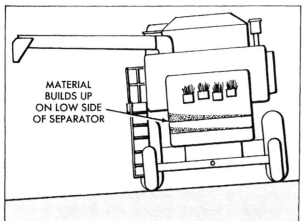

Fig. 14—Level-Land Combine Separator Tilts With The Contour Of The Ground

The separating and cleaning units can handle some degree of tilt, but if the combine becomes tilted too much, the crop being harvested builds up on the low side (downhill side) of the machine (Fig. 14). This will result in poor separating and cleaning action; the material may choke the machine or be pushed out the rear of the combine with little or no separating action. This, of course, causes crop losses.

Most manufacturers make available sidehill attachments for level-land combines which operate frequently on rolling ground. These attachments are usually shields or deflectors which prevent the material from falling to one side of the separator (Fig. 15). However, these devices can only reduce the material buildup; separating and cleaning action may still be affected when the combine operates on steep slopes.

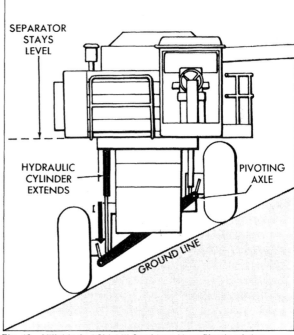

Fig. 16—Hillside And Sidehill Combines Have Pivoting Axles

Hillside and Sidehill Combines

The hillside combine is supported by pivoting axles which adjust to the changing slopes of hillsides (Fig. 16). The separator levels automatically on grades up to 45 percent on hillside combines and 18 percent on

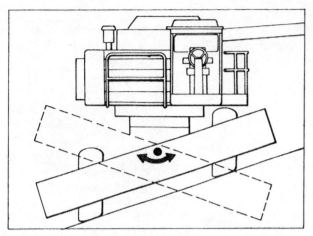

Fig. 17—The Platform Swivels Automatically To Conform To The Slope

Fig. 18—How is Size of Self-Propelled Combines Determined?

sidehill combines. By keeping the separator level, the most efficeint separating and cleaning action is maintained because material is distributed evenly throughout the separator — unlike the level-land combine separator which tilts with the slope of the land and causes a buildup of material on one side.

A combination of fluid-level, electrical, and hydraulic power systems make up the self-leveling mechanism. Large hydraulic cylinders (Fig. 15), connected to the separator and axles, adjust the separator to keep it level. The rear axle is mounted on a pivot and adjusts freely to the slope of the ground. The cutting platform (header) swivels on a pivot also and conforms to the contour of the hill (Fig. 17). See Chapter 3 for a full explanation of the operation of the hillside combine.

COMBINE SIZES

How is the size of a self-propelled combine determined? The first factors that come to mind are power rating, header size and separator size (Fig. 18). But these are only rough measures of combine size because manufacturers don't usually make combine components which measure exactly like those of other manufacturers. What the manufacturers are actually trying to describe is the *capacity* of the combine.

Fig. 19—Typical Combine Engine

Capacity cannot be measured entirely in terms of dimensions of the combine or its horsepower. Yet all these elements contribute to the capacity.

Consider the following when trying to determine combine size and capacity:

1. Engine Power
2. Separator Width and Length (total square inches or m² of area)
3. Type of Threshing Cylinder or Rotor
4. Header Size
5. Grain-Tank Capacity

ENGINE POWER ranges from 70 horsepower (52 kW) to 270 horsepower (200 kW) or more. When comparing combine sizes, consider engine power in relation to threshing method as well as separator width and length. For example, if one combine has a smaller separator but a larger engine than another combine of nearly the same size, the machine with the smaller separator may actually have more capacity when operating in crops with high yields.

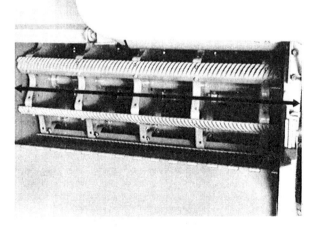

Fig. 20—Separator Width Affects Capacity Of Conventional Combines

SEPARATOR WIDTH AND LENGTH in conventional combines governs the maximum possible capacity of the combine to separate and clean the crop. This, of course, also depends on the efficiency of the design and the available power. Separator width varies from less than

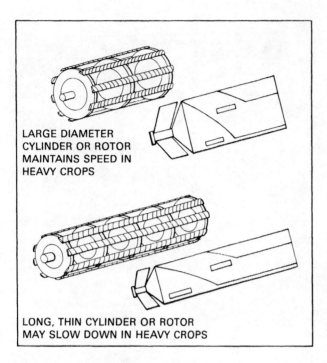

Fig 21—Type Of Threshing Cylinder Or Rotor Affects Combine Capacity

THRESHING CYLINDER OR ROTOR diameter can also affect capacity (Fig. 21). Small-diameter cylinders or rotors must rotate much faster to provide equal threshing action. They may lose momentum (ability to rotate at proper speed) more easily than larger-diameter cylinders or rotors when heavily loaded.

HEADER SIZE (Fig. 22) is governed by the size of separator and the power of the combine on which it is attached. For example, a 20-foot (6-m) header cannot operate properly on a combine with only 50 horsepower (37 kW). On the other hand, it would be inefficient to attach an 8-foot (2.4-m) header to a 150-horsepower (112-kW) combine. Headers are now available from 10 feet (3 m) wide to 30 feet (9 m) wide; corn heads vary from 2-row units up to 12-row units.

24 inches (0.61 m) to over 60 inches (1.52 m); separator length varies from about 105 inches (2.67 m) to 170 inches (4.32 m).

CAPACITY OF ROTARY COMBINES is affected by number, length, and diameter of rotors as well as total concave and separating grate areas. The design of the rotor feeding system as well as that of the rotor, concave, and separating system also helps determine combine capacity. Most rotary combines have only one rotor. But, some have two parallel rotors which turn in opposite directions. Rotor diameter ranges from 17 to 31.5 inches (0.43 to 0.80 m) and length from 88 to 168 inches (2.24 to 4.27 m).

Fig. 22—Typical Header With Pickup Reel

Fig. 23—Grain Tank Capacity is Important

GRAIN TANK CAPACITY usually corresponds with the overall size of the combine (Fig. 23). The capacities range from a little as 68 bushels (2.4 m³) on a 70-horsepower (52-kW) combine to more than 300 bushels (11 m³) on a 270-horsepower (200-kW) combine.

In summary, the capacity of each combine varies with the manufacturer, even though dimensions appear to be about the same. Determining the capacity of a combine is a matter of judgment, but the elements mentioned above must be considered individually and compared to those of other combines. Also, crop conditions and yields are factors which greatly affect the capacity of any combine.

PULL-TYPE COMBINES

The pull-type combine is drawn by a tractor (Fig. 24). The combine is powered by the power take-off shaft of the tractor or from an auxiliary engine mounted on the combine. Older, large pull-type combines required an auxiliary engine to operate the combine mechanism. The tractor only had to pull the weight of the combine.

Fig. 24—Typical Pull-Type Combine Drawn By A Tractor

With the introduction of larger and more powerful tractors, the auxiliary engine on pull-type combines has been eliminated. Now the tractor PTO powers the modern combine.

The previous discussion of combine sizes applies in general to pull-type combines, except that only a few sizes are available. The header averages about 13 feet (4 m) wide; separator width varies from about 44 inches (1.12 m) to 55 inches (1.40 m); separator length varies from 148 to 160 inches (3.76 m to 4.06 m); grain tank capacities range from 105 to 190 bushels (3.7 to 6.7 m^3). Tractor power requirements are about 80 PTO horsepower (60 kW) and up.

The gathering platform (header) on pull-type combine is offset to either the right or left of the separating mechanism, while the header for self-propelled machines is more nearly centered. The offset platform on pull-type combines is necessary because a hitch is required to attach to the tractor.

From the tractor seat, the operator can:

- **Engage or disengage the separator drive**
- **Engage or disengage the platform drive**
- **Raise or lower the platform**
- **Engage or disengage the unloading auger drive**

The operator must stop the tractor to make other operating adjustments on most combines, such as changing the threshing cylinder settings, fan speeds and cleaning shoe settings.

WINDROWING METHOD

When there are many weeds in the grain, when there is considerable moisture at harvest time, or when the crop ripens unevenly, it may be desirable to cut the grain with a windrower (Fig. 25) and thresh it later with the regular combine equipped with the windrow-pickup attachment.

The windrower, performing the function of the reel, cutterbar, and platform of the combine, cuts the grain and lays it on the stubble. An opening at the end or center of the platform (depending on the type of windrower

Fig. 25—Self-Propelled Windrower

being used) permits the cut grain to be laid on the stubble in windrow form.

When the grain is properly cured or when the moisture content is low enough, a special windrow-pickup platform is attached to the combine, or a pickup unit is attached to the regular combine platform (Fig. 26).

These attachments elevate the windrow onto the combine platform (Fig. 27). From this point on, the feeding, threshing, separating, cleaning and handling processes are the same as described for the regular combine.

In the northern part of the United States and the southern part of Canada, the growing season is usually too short for grain to ripen completely. Therefore, it is usually windrowed to speed drying so that it can be threshed and stored safely.

The windrow method of combine harvesting has extended the former boundaries of this method of harvesting. Many sections of the country where weeds or rainfall prevented use of the combine are now using the windrow method with remarkable success.

Fig. 26—Windrow-Pickup Attachment On Pull-Type Combine

Fig. 27—Windrow-Pickup Attachment in Operation

The windrow and pickup method are not used exclusively with the pull-type combine. Self-propelled combines also use this method of harvesting when conditions require it.

SPECIAL COMBINES

Many special combines have been developed through the years to harvest special crops. Two excellent examples are the *peanut combine* and the *castor bean combine* (Figs. 29 and 30).

Fig. 28—Self-Propelled Combine Using Windrow-Pickup Attachment

Fig. 29—Peanut Combine

Fig. 30—Castor Bean Combine

SPECIAL ATTACHMENTS FOR COMBINES

As harvesting grain with a combine has become more important, many special items of equipment have been developed to make the combine operator's task more productive.

Fig. 31—Special Attachments Make Combines More Adaptable And Productive Around The World

Combines may now be equipped with straw choppers or spreaders and shaft speed sensing units to warn the operator when the combine is malfunctioning. Signal devices that indicate when the combine is overloaded are also available. In addition, combines with heated and air conditioned operator's cabs can be purchased. These are a few of the special items that have been developed. Many of these attachments are discussed later in this book.

SUMMARY: COMBINE TYPES AND SIZES

1. Modern Combine Types are:

- **Self-Propelled Combines (engine provides power for both propulsion and separator)**
- **Pull-Type Combines (tractor pulled, power for separator is supplied by tractor PTO)**

2. Self-Propelled Combines may be:

- **Level-Land Combines**
- **Hillside Combines**
- **Sidehill Combines**
- **Special Combines**

3. Combine Sizes Are Determined By:

- **Engine Horsepower**
- **Separator Width and Length**
- **Type of Threshing Cylinder**
- **Header Size**
- **Grain Tank Capacity**

4. Combine Sizes Are A Rough Guide To Combine Capacity

5. Today's Pull-Type Combines Require Tractors of 80 Horsepower (60 kW) Or More

6. Windrow Combining Consists Of Harvesting A Previously Cut-And-Windrowed Crop.

7. Special Attachments Are Available To:
- **Chop Or Spread Straw**
- **Air Condition The Operator's Cab**
- **Warn The Operator Of Combine Malfunctions**
- **Adapt To Different Crops**

CHAPTER QUIZ

1. (Fill in blanks.) The name "combine" was developed for a machine that combines the jobs of _____ and _____ in one machine.

2. What are the two basic types of self-propelled combines?

3. Name the two ways combines may be classified according to the method of threshing.

4. Name three features of modern combines which make today's combine harvesting easier than it was 50 years ago.

5. List at least four areas which help determine the capacity and size of a combine.

6. Describe four things the operator can control from the tractor seat when using a pull-type combine.

7. Name three attachments available on today's combines.

8. Why are some crops cut and windrowed before combining?

2
Harvesting Systems

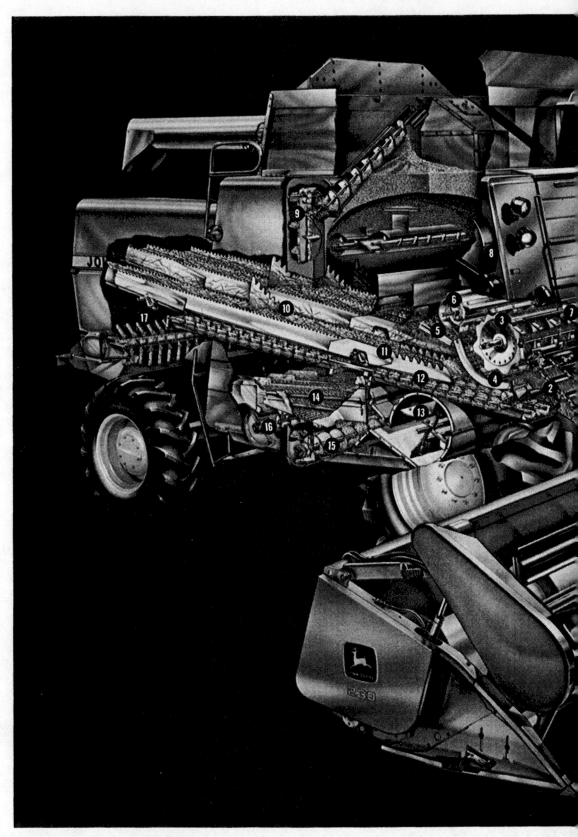

Fig. 1—Cutaway View of Combine In Operation

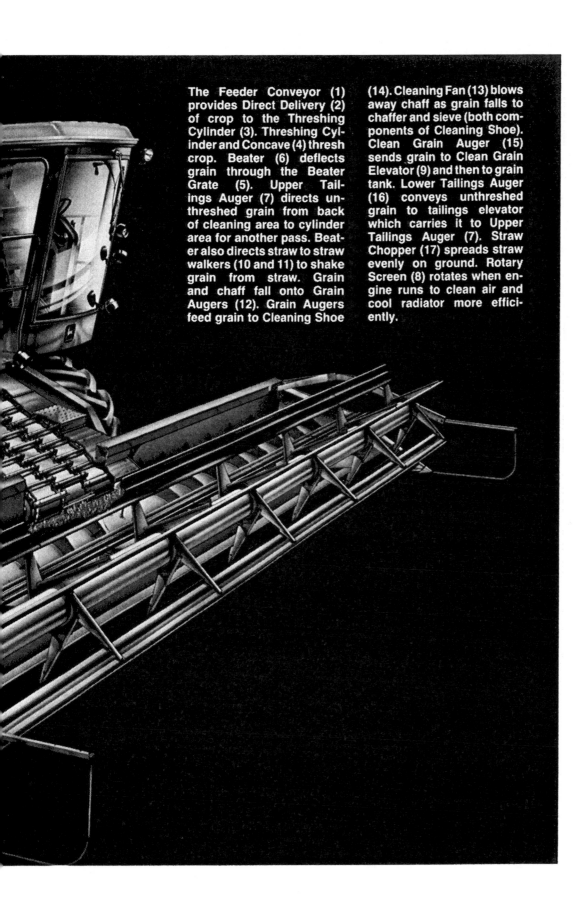

The Feeder Conveyor (1) provides Direct Delivery (2) of crop to the Threshing Cylinder (3). Threshing Cylinder and Concave (4) thresh crop. Beater (6) deflects grain through the Beater Grate (5). Upper Tailings Auger (7) directs unthreshed grain from back of cleaning area to cylinder area for another pass. Beater also directs straw to straw walkers (10 and 11) to shake grain from straw. Grain and chaff fall onto Grain Augers (12). Grain Augers feed grain to Cleaning Shoe (14). Cleaning Fan (13) blows away chaff as grain falls to chaffer and sieve (both components of Cleaning Shoe). Clean Grain Auger (15) sends grain to Clean Grain Elevator (9) and then to grain tank. Lower Tailings Auger (16) conveys unthreshed grain to tailings elevator which carries it to Upper Tailings Auger (7). Straw Chopper (17) spreads straw evenly on ground. Rotary Screen (8) rotates when engine runs to clean air and cool radiator more efficiently.

Fig. 2—Combine With Single Rotor And Concave

Fig. 3—Double Rotor And Concave Arrangement

Fig. 4—Transverse-Mounted Single Rotor and Concave Design

INTRODUCTION

Today's combine is a complex machine. Not only are the harvesting and threshing units complicated, but add to that the engine, power train, electrical system and hydraulic system, and it becomes one of the most complex machines in agriculture.

To understand the operation of a combine, look closely at each function of the machine. Once the operation of each of these components is understood, it becomes easier to understand how they relate to each other and the operation of the entire machine.

The basic operation of a combine is shown in Fig. 1. Figs. 1 through 4 also show basic differences in combine design. Fig. 1 is the conventional, cylinder-concave design while Figs. 2-4 are three variations of rotary combine designs. Cutting, cleaning, and handling functions are basically the same in all three models. The differences in threshing and separating functions will be described later in this chapter. This chapter describes each area of a typical combine.

All combines perform the following five basic crop harvesting functions:

1. **Cutting or Windrow Pick-Up and Feeding**
2. **Threshing**
3. **Separating**
4. **Cleaning**
5. **Handling**

CUTTING AND FEEDING THE CROP

The mechanism which cuts or gathers the crop and feeds it to the combine separator is commonly known as a header (Fig. 5). The header can be divided into distinct units: (1) the unit which cuts or gathers the crop may be a cutting platform, pickup platform, corn head or row-crop head; (2) the unit which feeds the cut or gathered crop to the separator is a feeder conveyor. See Figs. 6 and 7 to identify these parts. Each of these units will be discussed in detail later.

The header is attached to the combine by a pivot device which allows the header to be raised or lowered (by hydraulic cylinders) to obtain the desired height of cut (Fig. 8).

CUTTING PLATFORM OPERATION

Depending on the crop, a combine may be equipped with either a *regular cutting platform* which is used in most crops except corn and rice, or it may have a *"draper" platform* which is used in rice (Fig. 9). The draper platform is similar to the regular cutting platform except it has a "draper" or conveyor belt be-

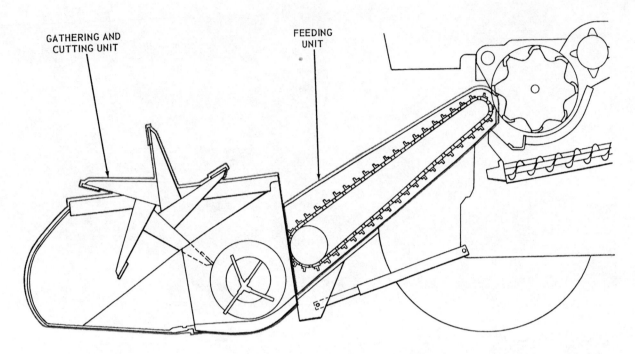

Fig. 5—Combine Header Unit

tween the cutterbar and the auger. The draper aids in picking up or getting more crop into the combine. Cutting platforms vary in width from 8 to 24 feet (2.40 to 7.20 m).

As the combine moves forward in the field (Fig. 10), the dividers and end sheets (1) separate a swath from the rest of the crop. The reel (2) parts a section of the crop and pushes it against the cutterbar (3). As the material is cut by the knife on the cutterbar, the reel continues to push the crop or lift it into the path of the spiralled auger (4). (On draper platforms, the reel lifts the crop onto the "draper" or conveyor which carries the material to the auger.) The auger moves the material to the center of the platform where the feeder conveyor (5) delivers it to the cylinder (6) for threshing.

The reel, cutterbar, auger and feeder conveyor must work in proper relationship to cut and feed the crop evenly to the threshing cylinder without losing the kernels or seeds.

Reel Operation

Two types of reels are available when combining various crops under different conditions:

Fig. 6—Typical Cutting Platform

Fig. 7—Typical Corn Head

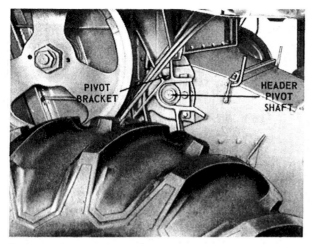

Fig. 8—Header Pivot Point

- **Bat-Type or Slat-Type Reel**
- **Pickup Reel**

The BAT OR SLAT REEL consists of three to eight slats made of wood or steel (Fig. 12). The slats rotate against the standing crop to hold it until the crop is cut by the knife on the cutterbar. Then the slat lays the crop back into the path of the auger.

The PICKUP REEL has several steel tines or "fingers" attached to the slats (Fig. 12). The fingers pick up crops which have been blown down or have become badly tangled, such as rice or barley. A slat-type reel without fingers cannot pick up crops in these conditions.

The fingers on the pickup reel reach down into the crop and lift it so the cutterbar can get under it. The pickup reel is also used in very ripe crops, such as soybeans, because a slat-type reel would knock the beans out of the pods, causing heavy crop losses. The fingers of the pickup reel tend to gather the ripe crop gently rather than batting it into the auger.

Both types of reels are usually adjustable. The slats on the slat-type reels and the fingers on pickup reels may be adjusted to enter the crop at the proper angle. This adjustment helps prevent shattering of the crop and permits uniform delivery of the crop to the platform auger. This adjustment is very important on a pickup reel which is being used in a down or tangled crop.

Reel Adjustment

Either type of reel can be adjusted fore and aft as well as up and down (Fig. 13). Both of these adjustments are important to insure proper delivery of material to the cutterbar and auger.

In standing grain the slat reel must be set so the slats, in their lowest position, strike just below the lowest grain heads and just slightly ahead of the cutterbar.

In crops that are down and tangled, the pickup reel must be set so it will pick up the crop and just clear the knife and conveying auger. This will insure that the material is picked up, cut off and swept back into the platform auger without losing grain.

The reel must also rotate at the proper speed to prevent shattering and loss of grain. Usually the speed for average conditions is 25 percent faster than ground travel speed. Attachments are available for some combines which automatically adjust reel speed in relation to combine travel speed.

The procedures for making reel adjustment will be described in Chapter 5, Field Operation.

Fig. 9—Two Types of Platforms

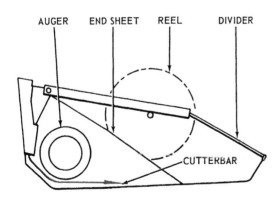

REGULAR CUTTING PLATFORM

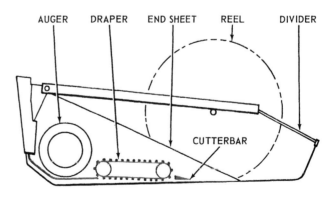

DRAPER PLATFORM

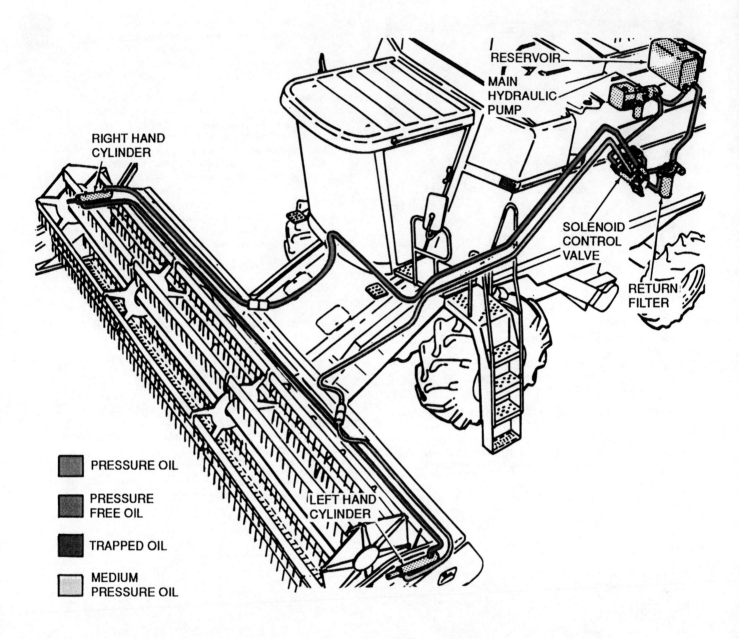

Fig. 10—Moving The Reel (Reel Forward Shown)

Reel Fore/Aft Operation

To move the reel forward or aft, the operator uses a switch on the armrest control panel. This energizes solenoids to move valve spools, allowing oil to flow.

When the solenoid control valve is in the neutral position, oil in the system is trapped. The trapped oil maintains a constant reel fore/aft position.

Pressure oil flows from main hydraulic pump, through the solenoid control valve, and to right or left hand cylinder. Trapped oil is transferred from the right or left-hand cylinder to the left or right-hand cylinder.

Medium pressure oil flows from left or right-hand cylinder to the solenoid control valve and then to the return filter and reservoir.

Cutterbar Operation

The reel holds the crop against the cutterbar as it is cut. The cutterbar consists of a steel angle bar on the front of the platform. Attached to the bar are stationary guards with slots through which the knife (sickle) moves back and forth, shearing or cutting the crop (Fig. 13). The knife is made up of several triangular blades (called knives or knife sections) which are riveted or bolted to a flat steel bar.

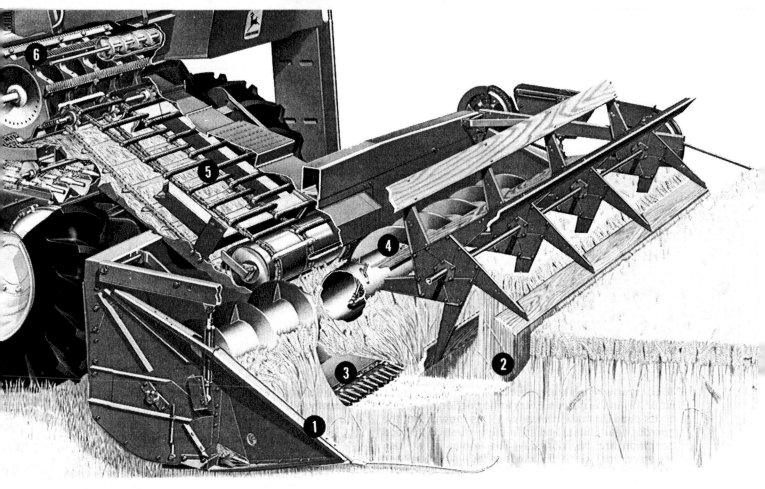

Fig. 11—Cutting Platform Operation

Fig. 12—Two Types of Reels

SLAT-TYPE REEL

PICKUP REEL

One end of the knife is connected to a reciprocating drive mechanism (Fig. 14) which causes the knife to move back and forth at several hundred strokes per minute. Some wide headers have two knives, each one half the length of the header, with a timed knife drive at each end of the header. This permits faster knife operation and reduces knife weight per drive unit.

Wear plates and knife hold-down clips control the relationship of the knives to the guards and complete the cutterbar assembly (Fig. 14).

Fig. 13—Reel Adjustment

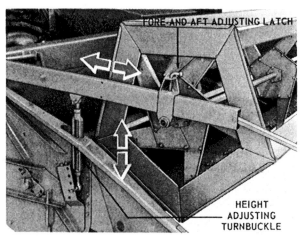

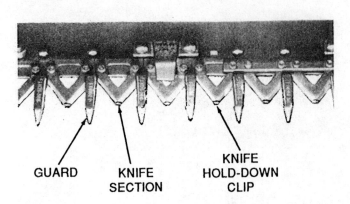

Fig. 14—Cutterbar

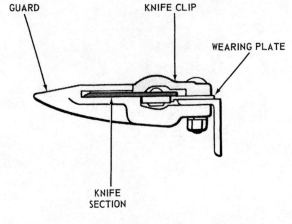

Fig. 16—Cutterbar in Cross-Section View

To cut properly, the knife must be sharp and must run smoothly in the guards. Every knife section must rest on its guard to make a *shear cut*. This means that the guards, wear plates, and knife clips must be in good condition and set correctly. The knives must also *register* properly with the guards as they move back and forth.

Failure to achieve a clean, shear cut, results in a tearing or chewing action at the cutterbar. This needlessly agitates a ripe crop, shattering kernels or seeds, and possibly results in large crop losses. It also causes increased cutterbar plugging.

A cross-seciton of the cutterbar is shown in Fig. 16. Notice how closely the milled slot conforms to the section. The section is about 1/8-inch (3.2 mm) thick and the slot is about 3/16-inch (4.8 mm) wide. This helps prevent material from wedging between the cutting edges.

Occasionally, when the cutterbar fails to operate properly, it must be adjusted; this adjustment is covered in Chapter 5, Field Operation.

Flexible Floating Cutterbar Operation

The same basic header and cutting parts are used on headers with either rigid or flexible cutterbars. However, instead of being rigidly attached to the front edge of the header, skid shoes and a special linkage permits the flexible cutterbar to closely follow ground contours (Fig. 17). This permits cutting much closer to the surface across the full width of the cutterbar. More seeds are saved in crops such as soybeans which usually have pods growing very close to the ground. The bar may be locked up for rigid operation in crops such as sorghum or small grains. These flex cutter bars usually flex about 4-in. (102 mm).

Automatic Header Height Control

Sensors under the combine header (Fig. 18) control height of cut when operating very close to the ground.

Fig. 15—Reciprocating Knife Drive Mechanism

Fig. 17—Flexible Floating Cutterbar

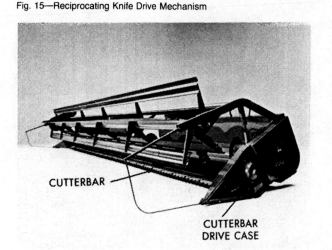

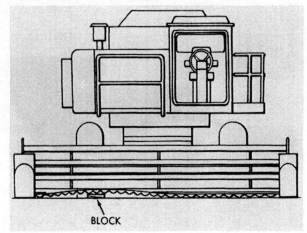

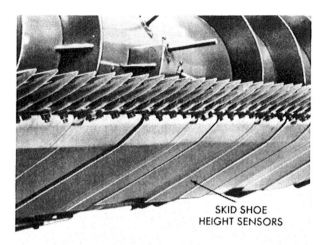

Fig. 18—Automatic Header Height Control

Fig. 20—Pickup Platform in Operation

for instance in soybeans. This relieves the operator of constantly adjusting header height and is particularly valuable when working over uneven ground or harvesting at night. The sensors may be adjusted to provide the desired height of cut under different harvesting conditions. Automatic header height control is standard on some flexible cutterbar headers and is optional on other models. Refer to the operator's manual for specific operating instructions.

Draper Operation

Drapers (conveyor belts) are used on special platforms for harvesting rice (see Fig. 9). Because rice is normally a "down" crop when harvested, gathering, cutting and feeding are difficult to do with a regular cutting platform. The draper platform uses a pickup reel which actually picks up the rice with steel fingers so that the cutterbar can cut it. After the rice is cut, the reel deposits it onto the drapers, which feed the material to the platform auger.

Fig. 19—Platform Auger

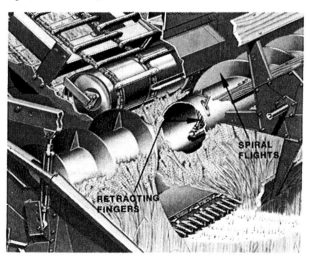

Platform Auger Operation

After the crop is cut by the cutterbar, the reel lays the material back onto the floor of the platform. Here spiral flights of the platform auger sweep the crop toward the center of the platform were the feeder conveyor is located (Fig. 19). Most augers have retracting fingers which move the material to the feeder conveyor for feeding the threshing cylinder. (Augers on draper platforms do not have these fingers.)

Smooth, even feeding of the material by the auger is essential for proper delivery to the feeder conveyor. The auger is adjustable to achieve correct feeding in most crops and conditions. This adjustment is covered in Chapter 5, Field Operation.

This concludes the basic information on operation of the cutting platform; next we will explain the operation of the pickup platform.

PICKUP PLATFORM OPERATION

Pickup platforms are used to gather crops which have been cut and gathered into a windrow by a grain windrower. The pickup platform is essentially the same as a cutting platform except it does not have a reel nor is a cutterbar required; it is equipped with a conveyor belt to which steel or plastic fingers are attached. These fingers pick up the windrowed crop and deliver it to the platform auger (Fig. 20).

A cutting platform may be converted to a pickup platform by installing a belt-pickup attachment.

CORN HEAD OPERATION

The corn head is essentially a corn picker mounted to the feeder conveyor of the combine. Corn heads vary in size from two-row units up to twelve-row units.

As the combine moves through the field, the gatherer points (1) are positions between the rows of corn

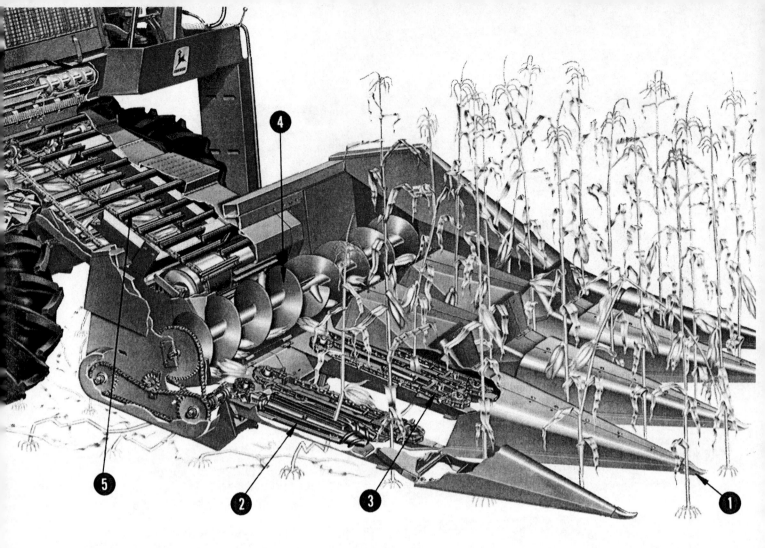

Fig. 21—Corn Head in Operation

(Fig. 21). Snapping rolls (2) grab the corn stalks and pull them rapidly down between the rolls.

When an ear of corn reaches the snapping bar, the ear is prevented from going through because of the narrow opening (Fig. 22). The snapping rolls continue to pull on the stalk and snap the ear free of the stalk.

Gatherer chains (3, Fig. 21) catch the ears and carry them to a cross auger (4) which delivers the ears to the feeder conveyor (5). The feeder conveyor delivers the ears to the threshing cylinder.

The snapping rolls must operate at a speed in direct relationship to the forward speed of the combine in order to pull the stalks through the rolls before the combine runs over them. If the snapping rolls are operated too fast, the ears may bounce off the corn head and be lost on the ground. High speed may cause shelling at the snapping bars which will also result in losses. It may also cause the entire stalk to be pulled into the combine, resulting in overloading the machine.

SNAPPING BARS AND GATHERER CHAINS

SNAPPING ROLLS (BOTTOM VIEW)

Fig. 22—Gathering Mechanism

Fig. 23—Row-Crop Head

If the speed is too slow, the ears will be snapped at the back of the rolls, causing congestion and possible plugging of the head.

The corn head must operate close to the ground to pick up low ears. The gatherer points are adjustable for various crop or field conditions.

In Chapter 5, Field Operation, adjustments of the corn head are covered.

ROW-CROP HEAD OPERATION

Row-crop heads (Fig. 23) are used in crops such as soybeans, grain sorghum, sunflowers, and similar crops to reduce cutting and gathering losses. The gathering points are very low and sloping so that they lift and guide lodged or tangled crops into the gathering area.

Rubber gathering belts on each row unit hold the plants as they are cut off by a knife. The knives are located just below the forward end of each set of gathering belts.

After the plants are cut, the gathering belts convey the crop to the cross auger at the rear of the header.

Each row unit is free to float up and down within a preset range to follow ground contour. This feature helps permit cutting of soybeans that grow close to the ground.

By eliminating a reciprocating knife and rotating reel, seed shattering is reduced. This permits faster operating speed than would be possible with a regular or flexible cutterbar header.

FEEDER CONVEYOR OPERATION

The movement of the material from the platform or corn head is done by the feeder conveyor chain or the rake in the feeder conveyor unit. The feeder conveyor takes the material from the platform or corn head and feeds it to the threshing cylinder.

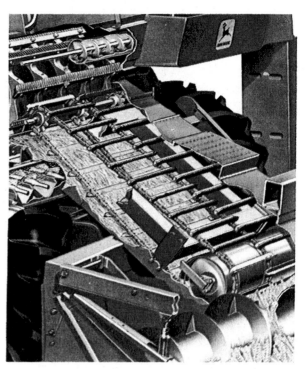

Fig. 24—Feeder Conveyor

In some combines, a feeder beater is used to help move the material from the platform auger to the feeder conveyor. This beater consists of a round drum equipped with retracting fingers which are similar to the fingers used in the platform auger. In one type of feeder conveyor, the feeder beater feeds the material from the platform directly to the cylinder. Some combines may have a series of paddles that move the material to the cylinder. The most common type of feeder conveyor consists of the chain conveyor or a combination of a feeder beater and chain conveyor.

The chain-type feeder conveyor is allowed to "float" at the lower end to permit smooth feeding of both small and large masses of material (Fig. 24). The conveyor chain is adjustable for various crop conditions. Adjustment is covered in Chapter 5, Field Operation.

A variable-speed feeder control is available for some combines which permits the operator to adjust feeder speed to crop conditions and travel speed. Some combines also have a reversible conveyor drive to aid in clearing the feeder conveyor in case it becomes plugged.

STONE PROTECTION

Most combines provide some form of protection from stones and other objects which might enter the machine and damage the cylinder, rotor, or concave. Some stone doors open automatically (Fig. 25) when a stone or some other solid object large enough to cause

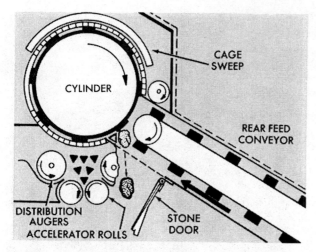

Fig. 25—Automatic Stone Door

Fig. 27—Threshing Section of a Conventional Combine

damage approaches the cylinder or rotor. When the door opens, a monitor light and alarm are activated and the operator must stop the machine to close the door. These stone doors rely on impact force of the object striking the door. This force is created by the object contacting the cylinder, rotor, or a special beater. Before closing the door, inspect these moving parts plus the feeder conveyor for possible damage. Make necessary repairs before resuming operation.

Successful operation of an automatic stone door requires proper adjustment and lubrication of the latch. This ensures that the door opens when needed and will not open unnecessarily. The monitor control must also be checked regularly so the operator is notified when the door opens to avoid heavy crop loss through the opening.

A cavity in front of the cylinder on some combines (Fig. 26) catches and holds stones and other hard objects. Depending on combine design, these traps may open and empty while the combine is operating or they may require the machine to be stopped and the door opened manually from under the feeder house.

Fig. 26—Stone Trap Must Be Opened Manually

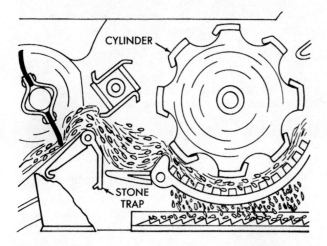

CAUTION: Be sure header lift cylinder safety stops are in place and the engine stopped before working under the header or feeder house.

During normal combine operation these stone traps fill with straw, grain, and dirt. The impact of an object after striking the cylinder normally causes it to displace enough material from the trap to lodge there rather than pass through the cylinder. However, if the trap is not cleaned regularly, the buildup of material may become so compacted that stones cannot lodge in the material. Consequently the stones are carried into the cylinder and concave anyway. Therefore, to maintain protection, empty the stone trap regularly whether stones are present or not.

THRESHING THE CROP

The "heart" of any combine is the threshing section (Figs. 27, 28). Webster's dictionary defines "threshing" as beating the grain from its husk or beating the grains from the heads, as in wheat. In the case of corn, this would be removing kernels from the cob, and in soybeans it would be removing the beans from the pod. In the threshing section of the combine, all the grain is threshed from the stems or pods. Up to *90 percent* of the grain is separated from its stem, cob, or pod as it passes between the cylinder or rotor and the concave. This vital area affects the entire operation of the combine, because if proper threshing isn't achieved here, the combine will fail to do its job.

The threshing area of a combine is located in the body of the combine which is known as the "separator." The components which make up the threshing mechanism (Fig. 27) consist of a cylinder (1) and a concave (2), or rotor and concave (1, Fig. 28).

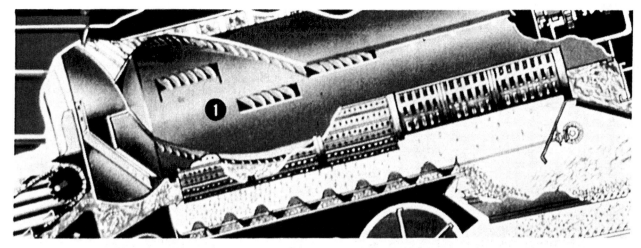
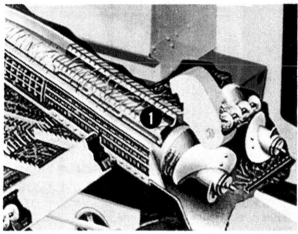
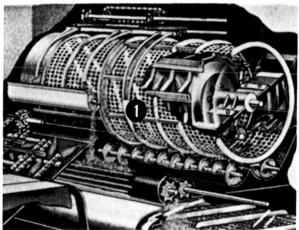

Fig. 28—Threshing Sections of Rotary Combines—Single Rotor (Top), Double Rotor (Left), Transverse-Mounted Rotor (Right)

TYPES OF THRESHING SYSTEMS

There are four basic types of threshing systems and matching concaves:

- Rasp-Bar Cylinder and Concave
- Spike-Tooth Cylinder and Concave
- Angle-Bar Cylinder and Concave
- Single or Dual Rotor and Concave

The most common design now is the *rasp-bar cylinder and concave* because almost all crops can be threshed with it. The *spike-tooth* design is used almost exclusively in rice or edible beans. The *angle-bar* design is found in a few machines and is used mostly for small-seed crops such as clover and alfalfa. Cylinder sizes may vary from 15 to 22 inches (38 to 56 cm) in diameter and from 24 to 60 inches (61 to 152 cm) in length.

Rotary combines with rasp bar rotors and matching concaves are now available. They can thresh essentially all the same crops harvested by combines with rasp bar cylinders. Rotors range from 17 to 31.5 inches (43 to 80 cm) in diameter and from 88 to 168 inches (2.24 to 4.27 m) in length. Both single and dual rotor models are available.

Here are more detailed discussions of each type.

RASP-BAR CYLINDER AND CONCAVE

The *rasp-bar cylinder* (Fig. 29) consists of a number of corrugated steel bars attached to the outer circumference of a series of hubs (Fig. 29). These hubs are mounted on a cross shaft at the front of the separator. The shaft is carried on ball bearings.

The cylinder is driven at speeds ranging from 150 to 1500 rpm. This variety of speeds is necessary to meet the different threshing actions required for various crops and crop conditions. Most crops can be threshed at speeds between 400 and 1200 rpm.

Fig. 29—Typical Rasp-Bar Cylinder

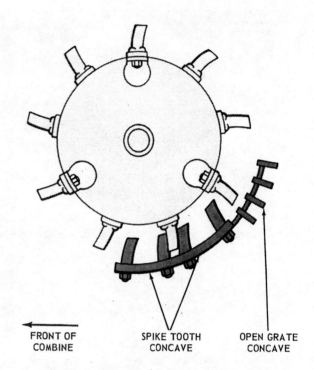

FRONT OF COMBINE — SPIKE TOOTH CONCAVE — OPEN GRATE CONCAVE

Fig. 31—Spike-Tooth Concave

The *concave* consists of a series of parallel steel bars held together by curved side bars and rods (Fig. 29). The concave is mounted under and slightly to the rear of the cylinder. The curvature of the concave generally conforms to the outer circumference of the cylinder.

The rasp bars have corrugations running in opposite directions on adjacent bars (diagonally). Usually, the direction of the corrugations are referred to as right or left to identify a particular bar. These corrugations provide rubbing or rasping action on the crop as it passes through the threshing area.

The rasp-bar cylinder and concave will handle damp weeds with less tearing of the stems than a spike-tooth cylinder. This means drier and cleaner grain in the tank and less overloading of the cleaning shoe.

SPIKE-TOOTH CYLINDER AND CONCAVE

The *spike-tooth cylinder* consists of a number of steel teeth attached to metal bars which are mounted to the outer circumference of a series of hubs (Fig. 30). The basic mounting and drive configuration is the same as described for the rasp-bar cylinder.

The *concave* also has teeth (Fig. 31). The teeth are attached to bars which are held in place by curved side bars. As with the rasp-bar concave, this concave is mounted under and slightly to the rear of the cylinder.

The teeth in this design tear and shred the material rather than rub and flail it as does a rasp-bar design. As the cylinder rotates, its teeth pass between the stationary teeth of the concave which causes the threshing action (Fig. 32).

The spike-tooth cylinder-concave unit is more aggressive than other types, and it will grasp and "digest" heavier volumes of material. Almost all rice combines are equipped with spike-tooth cylinders for this reason. Spike-tooth cylinders are also popular for edible beans because this cylinder has excellent threshing action with little seed damage.

The disadvantage of the spike-tooth design is that it tends to tear up the straw and weeds, which can result in separating and cleaning problems.

Fig. 30—Typical Spike-Tooth Cylinder

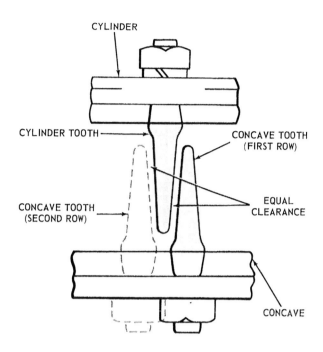

Fig. 32—Spike Cylinder Teeth Pass Between Concave Teeth

ANGLE-BAR CYLINDER AND CONCAVE

The *angle-bar cylinder* consists of helically mounted angle-iron bars attached to hubs (Fig. 33). Both the bars and the *concave* have rubber faces. The basic mounting and drive configuration of this design is the same as the others described above.

This design flails the grain rather than rubbing and flailing as does the rasp-bar type. The threshing action is considerably more gentle than either the rasp-bar or spike-tooth cylinder. It is commonly used for small-seed crops, such as clover or alfalfa.

Fig. 33—Typical Angle-Bar Cylinder and Concave

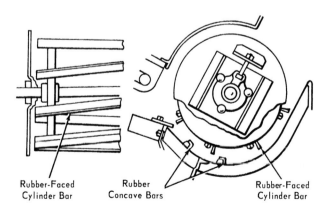

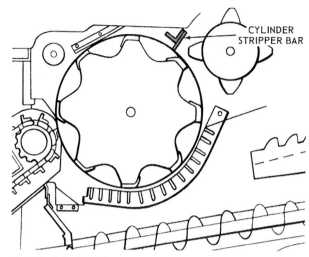

Fig. 34—Cylinder Stripper

CYLINDER STRIPPER

Most combines are equipped with a cylinder stripper (Fig. 34). The cylinder stripper is usually located near the top of the cylinder and is used to prevent backfeeding. Backfeeding occurs when material is carried around the circumference of the cylinder and dropped in front of the cylinder again.

This can seriously hamper the capacity of a combine. When a combine has backfeeding problems, it may have to travel slower than the same combine without the problem.

The stripper is usually adjustable. However, some manufacturers fix the stripper in a preset position and no adjustment is necessary.

ROTOR AND CONCAVE

Rasp bars, similar to those used on rasp bar cylinders, are attached to the input end of the cylindrical rotor in rotary combines. A matching concave under the rasp-bar portion of the rotor permits separation of up to 90 percent of the grain from the straw at that point. Rotary combines may have a single rotor, or two parallel rotors. Most rotors are mounted longitudinally in the combine and are fed from the front end of the rotor. Some models have a rotor mounted transversely (crosswise) which is fed from the side at one end of the rotor, much like a conventional cylinder. Typical rotor configurations are shown in Fig. 35.

Rotor speeds range from about 200 to almost 1600 rpm, depending on machine design and the threshing requirements of a variety of crops. Variable speed controls permit adjustment of rotor speed to meet changing crop conditions as the machine moves through the field.

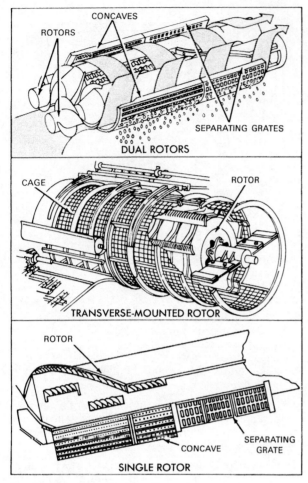

Fig. 35—Typical Rotor Configurations

Unlike the cylinder-concave design of conventional combines, the crop is exposed to more than one pass between the rotor and concave. Therefore, the clearance between rotor and concave is usually somewhat wider than conventional cylinder-to-concave spacing in the same crop and conditions. This wider spacing may result in less grain damage than is caused by conventional cylinders, particularly in high moisture crops.

THRESHING CYLINDER OR ROTOR AND CONCAVE FUNCTION

The feeder conveyor delivers the crop to the threshing area of the combine. The crop is fed into the opening between the cylinder or rotor and concave. As the cylinder rotates, the material comes into contact with the rapidly moving rasp bars and this impact shatters the grain or seed from the stem, cob or pod. Additional threshing occurs by a rubbing action as the material is accelerated and passes through the restriction between the cylinder or rotor and the concave (Fig. 36).

The impact and rubbing action of the rotor and concave on the crop are essentially the same in both rotary and conventional combines. And, regardless of the machine type, the speed at which rasp bars travel must be about the same for the same crop and conditions. For instance, corn usually requires an impact speed of about 3000 feet (914 m) per minute. This means that a 17 inch (43 cm) diameter rotor would need to operate at approximately 675 rpm, compared with 520 rpm for a 22 inch (56 cm) conventional cylinder to provide equal rasp bar impact.

Up to 90 percent of the seeds which have been broken loose by impact or rubbing are separated from the straw through the concave openings or grates. The amount of separation that takes place here directly affects the overall capacity of the combine. For instance, if very little separation occurs through the concave, more of the grain is thrown onto the straw walkers or straw racks which causes the limit of the separating capacity of these units to be reached much faster. This means that a higher percentage of grain will be lost at the rear of the combine at a given ground speed.

Fig. 36—Threshing Action

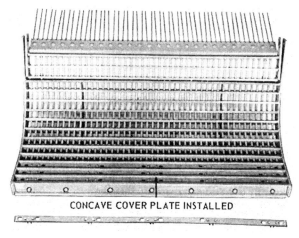

Fig. 37—Concave Cover Plates

For hard to thresh crops, concave cover plates can be used to increase the threshing action (Fig. 37).

On some rotary combines, concave filler wires must be installed when threshing small grain and other small-seed crops, and removed again when threshing large-seed crops such as corn and soybeans.

Combines with no provision for separation at the concave have other devices such as chain separators ahead of the straw walkers.

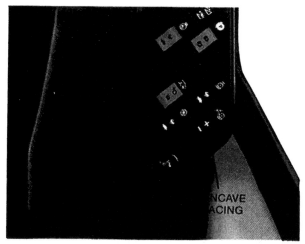

Fig. 38—Concave-To-Cylinder Spacing Adjustment

THRESHING CYLINDER OR ROTOR AND CONCAVE ADJUSTMENTS

Two basic adjustments are provided for the threshing cylinder or rotor and concave. These are:

- **Cylinder or Rotor Speed**
- **Concave Spacing**

Making these adjustments properly is vital for harvesting high-quality seed with a minimum of losses.

CYLINDER OR ROTOR SPEED

Cylinder or rotor speed basically affects two things:

(1) How many seeds are threshed from the straw, cob or pods

(2) How many seeds are broken or damaged in threshing

The so-called easy-to-thresh crops are those seeds which require a low impact to thresh them but can tolerate a high impact before they are damaged. When threshing this type of crop, speed adjustment is not very critical. But, when the impact required to thresh the crop is almost the same as the impact which will break the seed, then the speed must be carefully adjusted to get the most complete threshing with the least damage.

Cylinder or rotor speed is usually controlled from the operator's platform.

CONCAVE SPACING

Concave-to-cylinder or -rotor spacing is controlled from the operator's platform (Fig. 38). On most combines, including most rotary models, the concave is moved up or down as desired to get the proper spacing (Fig. 39). On other combines, the cylinder is moved up or down for this spacing; on these models the operator must make this adjustment with a wrench. The concave on

Fig. 39—Two Methods of Adjustment

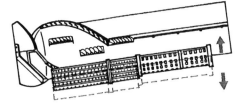

CONCAVE TYPE ADJUSTMENT
(Rotary Combine)

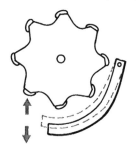

CONCAVE TYPE ADJUSTMENT
(Only Concave Moves Up and Down)

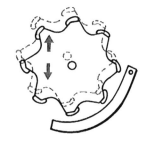

CYLINDER TYPE ADJUSTMENT
(Only Cylinder Moves Up and Down)

Concave spacing affects:

(1) How well seeds are threshed

(2) How many seeds are separated from the straw through the concave grate

When threshing is difficult, spacing is reduced to make the band of straw between the cylinder or rotor and concave *thinner,* causing more of the seed heads to come in contact with the rasp bars. A narrow spacing between the concave and cylinder or rotor can also result in more grain being separated from the straw (Fig. 41). Seeds will move through the *thinner mat* of straw with the narrow spacing much more readily than through the *thicker mats* at the wide spacings. With proper clearance, more threshing occurs at the front of the concave and more of the seed falls through the grate before it is discharged from the concave area. With wide concave spacing, threshing occurs farther toward the rear of the concave and the grain does not have time to be separated through the concave.

As we mentioned earlier, the concave is usually adjusted so that the straw entry end is spaced farther from the cylinder or rotor than the discharge end. This

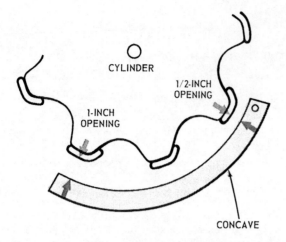

Fig. 40—Typical Front and Rear Spacing Proportions

some rotary combines is adjusted with a ratchet on the side of the combine. Most combines have adjustments for the front and rear of the concave. Always refer to the operator's manual to determine the proper front and rear spacing. Spacing at the front of the concave may vary from a fraction of an inch (or centimeter) to 1-1/2 inches (3.8 cm). The spacing at the rear of the concave is affected by the spacing at the front. Normally the rear spacing is about one-half the dimension selected at the front (Fig. 40). On rotary combines, the rotor-to-concave spacing may be the same along the full length of the concave. Most on-the-go adjustments change front and rear spacing at the same time and maintain the same clearance ratio through the full range of adjustment.

Fig. 41—Cylinder-Concave Spacing for Small Grains

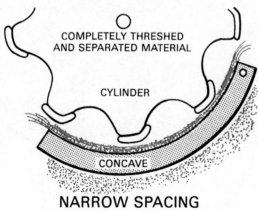

**NARROW SPACING
BETTER THRESHING ACTION**

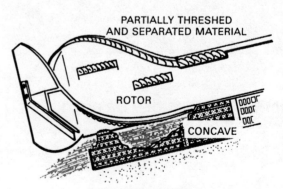

**WIDE SPACING
POOR THRESHING ACTION**

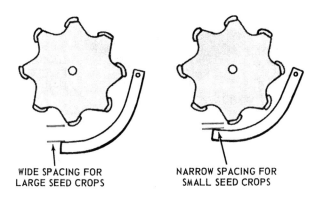

WIDE SPACING FOR LARGE SEED CROPS

NARROW SPACING FOR SMALL SEED CROPS

Fig. 42—Concave-Cylinder Positions For Small- and Large-Seed Crops

gives a *wedge* or *funnel* effect and helps material to feed uniformly and without hesitation into and through the threshing area.

To achieve the optimum operation of the cylinder and concave, the components must be adjusted to-

Fig. 43—Concave Must Be Parallel With The Cylinder

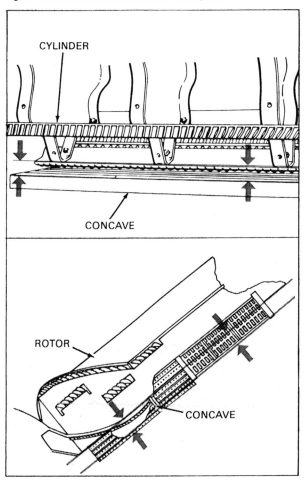

gether (Fig. 42). When harvesting *small-grain or small-seed crops* such as wheat, barley, rye, maize, alfalfa, clover, etc., the concave spacing is adjusted so that the crop is completely removed from the straw or stem without excessively tearing up the straw or stem. This is judged by the condition of the straw discharged by the combine.

If the concave is set too close for the crop being harvested, the straw will be excessively torn up and more horsepower will be required to thresh the crop.

If the concave is set too wide, the crop will not be completely threshed.

After the concave is properly adjusted, the cylinder or rotor speed is then adjusted to achieve maximum threshing with the least crop damage. If crop damage does occur, *do not widen the concave clearance. Instead, reduce the cylinder speed.* Concave spacing in these crops has very little effect on seed damage.

In the *large-seed crops* such as soybeans and corn, the same procedure holds true except that the concave spacing is somewhat wider (Fig. 42). In corn, this spacing is dictated by the condition of the cobs and ease of shelling. Usually this space is adjusted open far enough to keep the cobs whole or in as large pieces as possible during the threshing process.

If some varieties of corn are harvested early when the cob is soft, concave spacing must be closer for satisfactory shelling. Seeds of the large-seed crops usually are damaged very easily.

Because concave spacing is often critical, *be sure that the concave is kept parallel with the cylinder or rotor (Fig. 43).* This will help ensure uniform threshing across the complete width of the cylinder or the full rotor length.

Often where there is a problem of grain loss due to "carrying over" at the shoe, it can be corrected by leaving the shoe adjustments alone and going back to the cylinder or rotor to make needed corrections.

EFFECTS OF SEPARATOR SPEED AND CYLINDER (ROTOR) - CONCAVE SETTINGS

Regardless of the crop harvested, *the combine must be run up to speed.* Many operators will reduce or increase engine speed in an attempt to reduce cracking at the cylinder or rotor or to get better threshing of the crop. This completely upsets the balance of other units in the combine.

Reducing overall speed also reduces speed of platform, straw walkers, shoe, and elevators. This sluggishness can result in clogging of the entire combine and increased grain losses.

Increasing overall speed will cause material to pass through the machine too rapidly, resulting in grain losses and naturally more strain and wear on all moving parts.

Fig. 44—Examine These Areas To Determine Threshing Action

Never alter the basic recommended speed of the combine. Use the correct speed and concave setting to help ensure thorough threshing. Always refer to the combine operator's manual for the basic settings recommended. From there, slight adjustments both in speed and spacing may be necessary to correspond with the crop and crop conditions.

Remember — All of the crop is threshed at the cylinder or rotor and 90 percent can be separated here.

Ideally, the cylinder or rotor and concave should be adjusted as follows:

1. *In small grain and beans,* adjust to remove all grain from the stalk with no damage to the grain, and to produce no broken straw or chaff so separation and cleaning are more efficient.

2. *In corn,* adjust to remove all kernels from the cob, with no damage to the kernels, and to pass as many whole cobs through as possible.

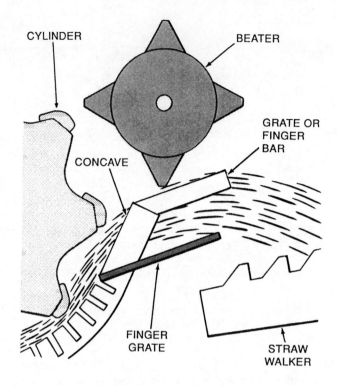

Fig. 45—Beater and Finger Grate

It is nearly impossible to achieve perfect performance. However, *by adjusting to near the point of under-threshing,* very satisfactory performance may be obtained. Adjustments to the cylinder or rotor and concave are discussed in more detail later in Chapter 5.

DETERMINING THRESHING ACTION

The only method of determining threshing action is by examining the material in the grain tank and tailings elevator, and the straw discharged from the rear of the combine (Fig. 44).

Fig. 46—Separating Section Of A Conventional Combine

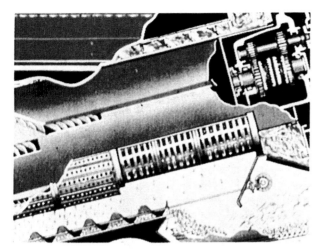

Fig. 47—Separating Section Of A Rotary Combine

SEPARATING THE CROP

The beater or rotary deflector is located directly behind and usually slightly above the threshing cylinder or rotor (Figs. 46, 47). It has a small diameter and about the same width as the threshing cylinder or rotor.

Up to 90 percent of the grain may be separated from the straw, cob, or husks as the crop is threshed by the cylinder or rotor. In conventional combines, the remaining loose grain is separated by the beater (1), finger grate (2) and straw walkers (3) shown in Fig. 46. The finger grate and straw walkers are eliminated from rotary combines, and final separation of grain and straw takes place as material flows through the rotor chamber and a beater (Fig. 47).

We'll look first at the separating process in conventional combines, then study how it is done in rotary combines.

CONVENTIONAL SEPARATION

Only loose kernels can be separated from the straw by the beater, finger grate and straw walkers — no threshing occurs in this area. Thus, unthreshed kernels will remain attached to the straw.

BEATER AND FINGER BAR OR GRATE

Four types of beaters or deflectors (Fig. 48) are:

- **Wing-Type**
- **Drum-Type with Removable Wings (Cover Teeth)**

The grain in the tank (1) will show the amount of damaged grain and foreign material. However, if kernel damage is excessive, it may be caused by too many threshed kernels being rethreshed from the tailing elevator (2) rather than improper adjustment of the cylinder and concave.

The straw discharged from the rear of the combine (3) should contain only an occasional low-quality kernel in the heads, and the straw should be as whole as possible.

These problems and how to correct them are discussed in Chapter 5 of this book.

Fig. 48—Four Types of Beaters or Rotary Deflectors

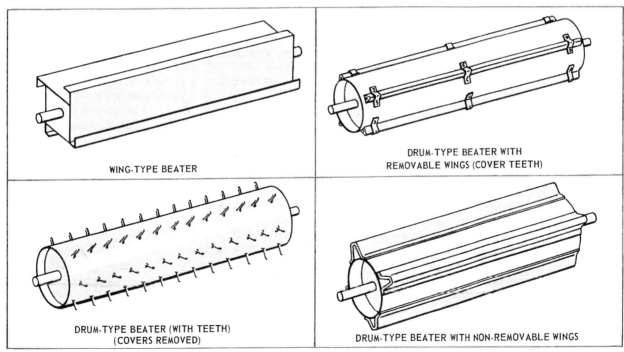

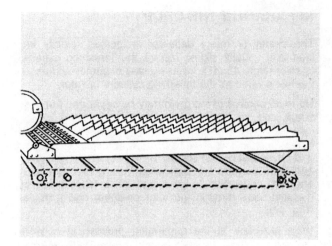

Fig. 49—Typical Straw Rack

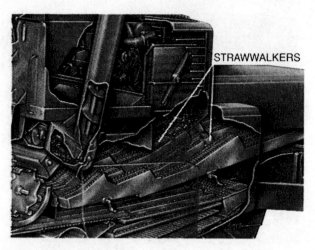

Fig. 50—Typical Straw Walker

- **Drum-Type with Teeth (Or Covers Removed)**
- **Drum-Type with Non-removable Wings**

The beater has two jobs:

(1) It slows material as it leaves the cylinder and concave.

(2) It deflects this material down onto the front of the straw walkers.

If the material to be separated is not deflected down to the extreme front of the straw walker surfaces, valuable separating area is lost.

The finger grate at the rear of the concave (Fig. 45) holds the material up so the beater can deflect it onto the straw walkers. Without the fingers, much of the material could fall down into the cleaning section, and overload it. The fingers also allow loose grain to fall through to the cleaning area.

STRAW WALKERS OR RACKS

Effective separation in a combine is determined by how the crop is shaken as it travels through the separating area. The straw walkers (or straw rack) not only provide the agitation to separate the remaining grain, but they

Fig. 51—Straw Walker Crankshaft

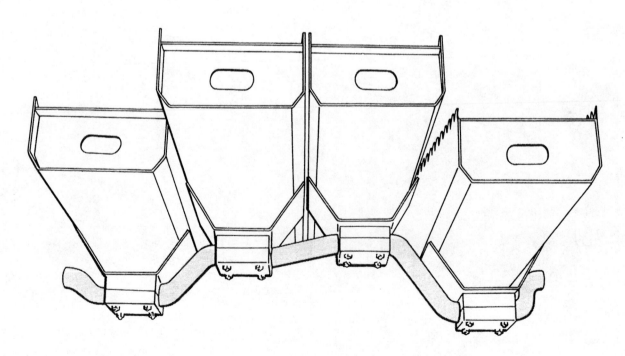

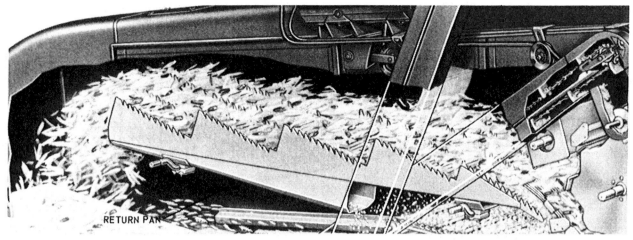

Fig. 52—Straw Walker With Return Pans

also remove the straw and trash by "walking" it out the rear of the combine.

The two most common types of straw carriers are the one-piece oscillating straw rack (Fig. 49) and the multiple straw walkers (Fig. 50). The straw rack fits across the width of the separator and is mounted so that it oscillates back and forth as a unit. The straw walkers are attached to multiple-throw crankshafts at the front and rear (rear crankshaft shown in Fig. 51). From three to five straw walkers are mounted in the separator, depending on the width of the combine. Each walker is positioned by the crankshaft at 90 or 120 degrees around the circle of crankshaft rotation.

Some straw walkers or racks have return pans under them which allow the grain to travel or roll forward down the pan to an opening located just over the front of the cleaning section (Fig. 52). Other designs are open along the entire bottom and the free grain can fall through onto a series of augers or a grain conveyor which moves the grain to the cleaning area (Fig. 53).

The most popular design of straw carrier today is the straw walker. Different types are used in different crop conditions. The most common straw walker is a step-type walker which provides excellent tumbling and walking action; two types are shown in Figs. 52 and 53. Sawtooth edges along each side of the walker help in agitating the straw as it moves through the combine.

Another walker, which is sometimes used in maize to eliminate plugging of heavy green stalk material, is called a one-step walker (Fig. 54).

Straw walkers have holes of different shapes and sizes to allow grain to fall through yet prevent straw and trash from going through. For many years a square or rectangular opening (Fig. 55) had been used. But because much corn is now harvested with combines, a "hooded" or "lip-type" opening has been developed to reduce plugging by stalks and cobs (Fig. 55).

Fig. 53—Straw Walker With Open Bottoms

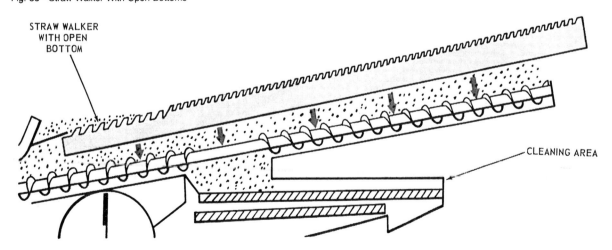

45

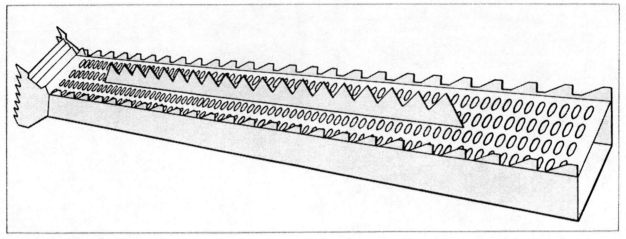

Fig. 54—One-Step Straw Walker

For some crop conditions, "fishbacks" (Fig. 56) are attached to the walkers to provide additional agitation and reduce the speed of the material traveling over the walker surfaces. However, if the straw is extremely fluffy or damp, the fishbacks may prevent material from moving away from the beater. If material is held too long at this point, the build-up will be picked up by the beater and cylinder and cause back feeding. If this happens, remove the fishbacks.

STRAW WALKER ACTION

After straw is deposited onto the straw walkers, it is tumbled and tossed as it is propelled to the rear (Fig. 57). The loose grain falls down through the openings in the walkers or rack and is carried to the cleaning shoe. The straw continues to be agitated along the length of the straw walkers until it reaches the rear of the combine and falls to the ground.

The straw walkers throw the straw in an upward and rearward direction during a portion of the agitation cycle. This leaves the straw momentarily in mid-air. The material then falls onto a section of the walker nearer to the discharge end. Each cycle "walks" the straw a little farther toward the rear.

Each agitating cycle occurs 150 to 250 times per minute depending on the combine. If the speed is either too fast or too slow, grain losses may increase. Always refer to the operator's manual to determine the proper speed. The speed of the straw walkers or racks is usually not variable but is determined by the basic speed of the separator.

Some operators have the misconception that better separation can be achieved by speeding up the walkers, but actually this can result in poor separation. With increased speed, straw moves through the machine too quickly for all the grain to fall through the walker openings. This, of course, results in greater grain losses.

Most combines have *high clearance* or headroom above the straw walkers to insure that the tossing action in heavy crops is not hindered (Fig. 58).

Fig. 55—Hole Shapes in Straw Walkers

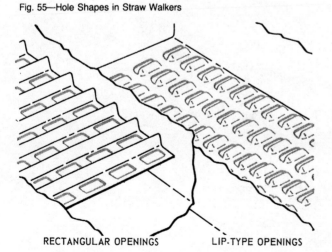

RECTANGULAR OPENINGS   LIP-TYPE OPENINGS

Fig. 56—Straw Walkers "Fishbacks"

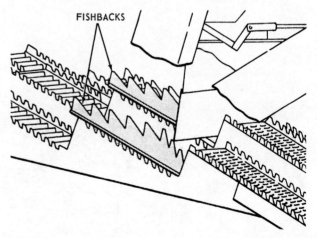

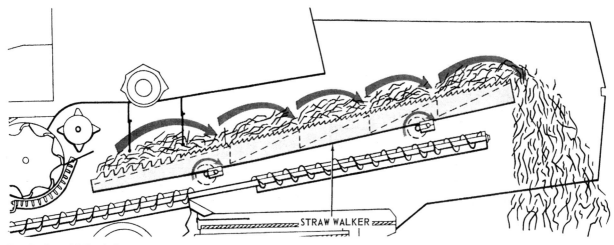

Fig. 57—Straw Walker Action

Curtains (or retarders) over the straw walkers for rack help to retard or slow down the flow of material, giving more time to agitate the straw and release grain or kernels (Fig. 59). They also help prevent the grain from being thrown by the cylinder over the walkers and out of the combine. These curtains are made of rubber or canvas material and go the full width of the combine separator. Usually one to three curtains are used. Curtains may be added or removed depending on the condition of the crop. If the crop is damp and there is difficulty in getting the grain to fall freely from the straw, additional curtains may be required to slow down the flow of material for better separation.

Fig. 58—High Clearance Is Needed Above Straw Walkers

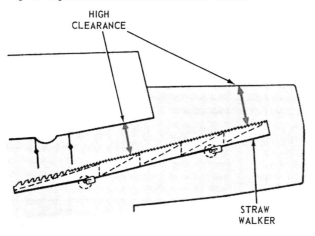

IMPORTANCE OF GOOD SEPARATING ACTION

The function of transporting the straw is a very important one. The straw must be moved through the machine quickly enough for good material-handling capacity, but not so fast that the grain cannot be separated from it.

As we mentioned before, if the straw is transported too slowly, the beater may catch the pile of straw and feed it back to the cylinder which will then carry it completely around and back through the concave (Fig. 59). This "backfeeding" is undesirable because the backfed material added to the new material entering the threshing cylinder causes overloading. Such an overload results in wasted power and plugged cylinders. Straw moving too slowly also results in a very deep mat of straw on the straw rack or straw walkers and reduces the amount of grain falling to the cleaning shoe.

In nearly all crops, the effective capacity of a combine is limited by the loss of free grain not separated from the straw before it is discharged from the combine. In other words, separation capacity of a combine is usually reached before its straw-handling capacity is reached. Combines are purposely designed this way to reduce the chances of plugging the combine when it is running at capacity.

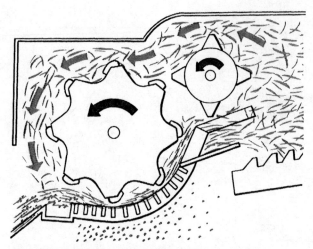

Fig. 59—Backfeeding Caused By Straw Build-Up

To obtain the maximum grain separation in a combine, the concave clearance and cylinder speed must be properly adjusted because most of the grain is separated by these components. Poor separation in this area cannot be completely overcome by the straw walkers.

The chart in Fig. 60 shows the typical separating loss versus feed rate characteristics of one combine model operating in wheat, rye or other similar crops.

For example, if this combine were to travel through the field at a speed that feeds straw and chaff in at the rate of 500 pounds (227 kg) per minute, 3 percent of the grain would not be separated from the straw. The capacity of this combine, in the particular crop that produced this curve, is from 400 to 500 pounds (181-227 kg) of straw per minute. As the straw input increases over this rate, the separating loss increases quite rapidly. This is typical of all combines because as the input of the straw is increased, a point is eventually reached at which the mat of straw flowing between the concave and cylinder becomes so compressed and dense that much less grain can be separated from the straw.

Also, the mat of straw on the walkers reaches such a depth that very little grain can be shaken out.

The "separation curve" for a given combine varies for the different crops as well as for conditions of the crop. However, all combines experience the characteristic rapid increase of losses beyond a certain feed rate. The amount of grain lost by the separating system is determined by the amount of straw fed through it. If a combine is operated at high travel speeds, it may handle the straw without plugging, but the separating loss will be too high. How to correct grain losses is covered in more detail in Chapter 5.

ROTARY SEPARATION

As a crop spirals through the rotor chamber, past the rotor rasp bars and concave, centrifugal force moves grain outward through the layer of straw and chaff to the outside of the chamber (Fig. 61). When grain reaches the separation grates around, or partially around the rotor, it passes through the grate, and straw continues to move toward the rotor discharge (yellow arrow). Curved fins or vanes attached inside the chamber act like rifling in a gun barrel to spiral straw and chaff around the chamber. Vanes on the rotor keep material agitated and moving toward the discharge. This aids separation by preventing straw or husks from matting and carrying grain through the combine.

As straw leaves the rotor chamber of most rotary combines a beater or impeller directs material toward the rear of the machine (Fig. 62). The action of the beater or impeller also helps remove any remaining grain from the straw and deflects it through a grate where it is carried to the chaffer.

As with conventional combines, overloading the machine makes it more difficult to obtain complete threshing and separation of grain and straw. Therefore, the feed rate or operating speed of rotary combines must be matched to crop density and condition to ensure satisfactory performance.

Overthreshing results in more damage to grain. Chewed up straw passes through the concave and

Fig. 60—Typical Separating Loss Versus Feed Rate

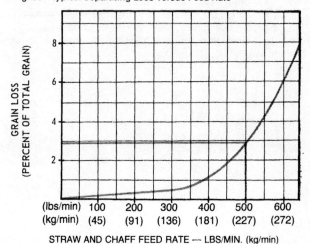

Fig. 61—Rotary Separation System

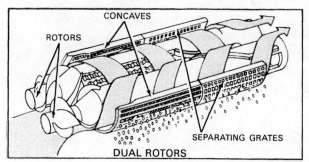

48

Fig. 62—Beater Removes Some Grain From Straw

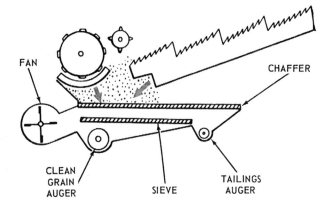

Fig. 63—Gravity Feed to Cleaning Unit

grates to overload the chaffer. Underthreshing can cause grain to pass through the combine, still attached to the head or cob. Therefore, proper concave and separation-grate clearance and rotor speed are major factors in determining the separating efficiency of rotary combines.

Separated grain leaving the rotor chamber is collected in an outer shell and falls directly onto the shoe or into a reciprocating grain pan under the separation grates. This grain is carried to the shoe for cleaning. On some machines, a series of parallel auger conveyors catch grain falling through the concave and from part of the separation grate, and deliver it to the shoe.

Because rotary combines rely on centrifugal separation, operation on sloping land doesn't cause material to pile up on the downhill side of the combine.

CLEANING THE CROP

After the crop is threshed and separated, the grain and chaff must be delivered to the cleaning area of the combine by gravity or a conveying system.

Four basic methods are used to deliver grain to the cleaning area:

- **Gravity Feed**
- **Conveyor Belts or Chains**
- **Multiple Augers**
- **Reciprocating Grain Pan**

A basic GRAVITY FEED design is shown in Fig. 63. Grain threshed by the cylinder and concave falls directly into the cleaning unit. Grain and chaff, separated from the straw, are returned to the cleaning area by the straw walker return pan.

Two basic CONVEYOR BELT OR CHAIN designs are shown in Fig. 64. In one design (B) the grain conveyor

Fig. 64—Conveyor Belts or Chains Deliver Grain to Cleaning Unit

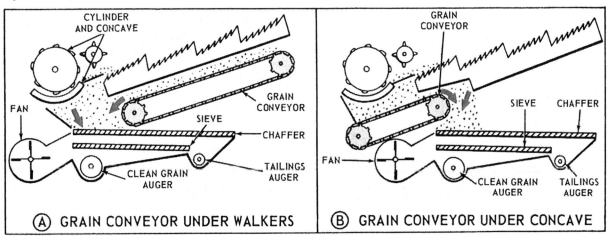

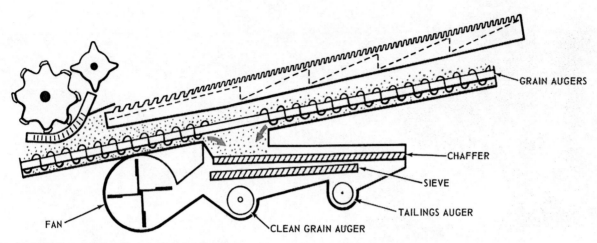

Fig. 65—Multiple Augers Deliver Grain to Cleaning Unit

is located under the cylinder and concave; the conveyor delivers threshed grain to the cleaning unit. The straw walkers have return pans which return the separated grain and chaff to the cleaning area. In design (A) the grain threshed by the cylinder and concave falls directly into the cleaning unit. The straw walkers are open on the bottom and separated grain and chaff fall onto the conveyor, (sometimes called raddle) which delivers it to the cleaning unit.

MULTIPLE AUGERS (Fig. 65) running the length of the threshing and separating areas move the grain to the cleaning unit. Grain separated by the cylinder and concave is moved back to the cleaning area by augers at the front. Augers at the rear of the separator have reverse flighting which pushes the grain separated by the straw walkers forward to the cleaning unit.

A stepped RECIPROCATING GRAIN PAN (Fig. 66) catches grain from the rotors of some combines. Shaking action of the pan separates material into layers of grain on the bottom, then chaff and small pieces of straw on top. Pan motion carries the material rearward and a finger grate on the end of the pan permits grain to drop directly onto the chaffer while air from the fan carries chaff and straw out the rear of the combine.

CLEANING UNIT

After threshing and separation, some chaff and straw are still mixed with the grain; the cleaning unit removes this trash from the grain. To do this, most combines have three basic components (Fig. 67) which make up the cleaning unit: a fan (1), a chaffer (2), and a sieve (3). The fan has its own housing; the chaffer and sieve are in the unit usually known as the "cleaning shoe".

These components will now be discussed in more detail.

Fig. 66—Reciprocating Grain Pan Delivers Grain to Chaffer

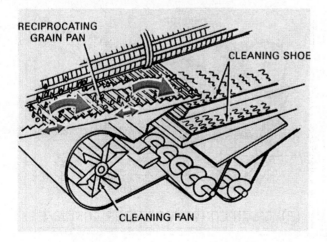

CLEANING FAN

The cleaning fan is a multiple-bladed fan mounted in front of the cleaning shoe (Fig. 68). The air blast from the fan removes most of the chaff and straw from the grain. Fan speeds may be adjusted from about 250 to 1500 rpm on most combines, depending on crops and conditions. However, on some combines, fan speed is constant and air flow is controlled by changing fan openings.

The amount of air can be controlled by three methods:

- **Shutters**
- **Windboards**
- **Fan speed**

Some cleaning fans are equipped with fan shutters and windboards (Fig. 69).

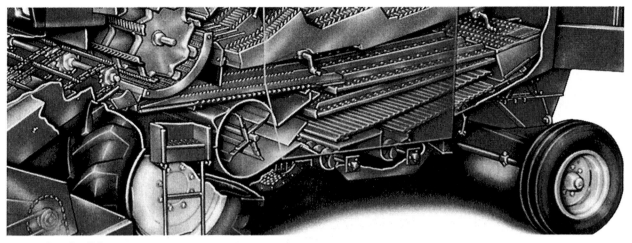

Fig. 67—Cleaning Unit

SHUTTERS are used to control the amount of air taken in and delivered by the fan. These shutters are partially closed when cleaning light seed and are wide open when cleaning heavy seed.

WINDBOARDS, located in the fan throat, control the direction of the fan blast to the chaffer and sieve. Usually the fan blast is directed well to the front of the cleaning shoe in heavy crops and directed more to the rear of the shoe in light crops. Adjustment of the windboards is critical because if the air blast is too far to the front of the shoe, material will accumulate at the rear of the chaffer and grain will be carried out of the combine. If the air blast is too far to the rear of the shoe, material will accumulate at the front of the chaffer and will result in poor cleaning action.

In many combines today, the fan housing and fan throat are designed to direct the air to the section of the shoe where it will do the most effective cleaning job. The need of windboards or side shutters has been eliminated by effective design and variable fan speeds.

FAN SPEED is adjusted along with the chaffer and sieve openings. In dry crops, the chaffer and sieve are opened more than normal for the crop being combined. This helps prevent the chaffer and sieve from riding grain out the rear of the combine or into the tailings. The fan is then adjusted for maximum usable air volume without blowing the grain into the tailings or out the rear of the combine. Slight adjustments are then made to get the most effective cleaning action. Problems of poor cleaning and grain losses caused by fan adjustments are covered in Chapter 5.

CLEANING SHOE

The cleaning shoe which contains the chaffer and sieve is a housing mounted under the main frame of the combine separator. The bottom of the shoe usually contains the lower tailings auger and the lower clean grain auger which will be discussed later in this chapter.

Fig. 68—Cleaning Fan

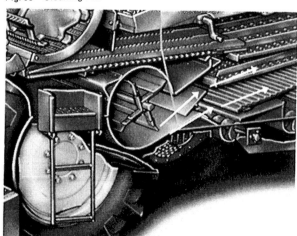

Fig. 69—Shutters and Windboards

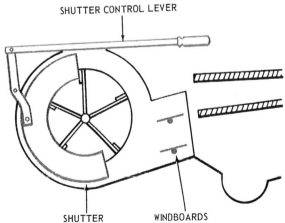

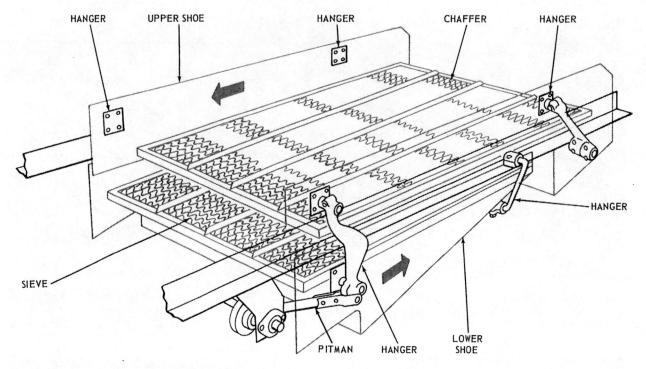

Fig. 70—Reciprocating Shoes Move Chaffer and Sieve

The chaffer and sieve are suspended on hangers, mounted on rubber bushings attached to the sides of the cleaning shoe (Fig. 70). The chaffer is mounted in the upper shoe (shown in blue); the sieve is mounted in the lower shoe (shown in red). The chaffer and sieve are moved back and forth by a pitman-type drive attached to the hangers.

Three types of typical shoe action are: reciprocating, shaker and cascading. In the "reciprocating shoe" the chaffer and sieve move in opposite directions to each other — opposed motion (Fig. 70). In the "shaker shoe" design the chaffer and sieve move in the same direction at the same time (Fig. 71). The third type is a "cascade shoe" which also uses chaffers and a sieve positioned so that the material drops from one unit to the other in a cascading or rolling motion as it is cleaned (Fig. 72).

Chaffer

Chaffers are available as either adjustable type or non-adjustable type. The *adjustable chaffer* is made up of a series of cross pieces of overlapping metal

Fig. 71—Shaker Shoe

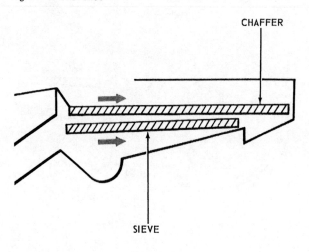

Fig. 72—Cascading Shoe

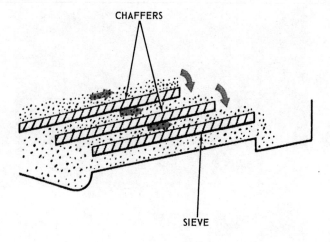

Fig. 73—Three Types of Adjustable Chaffers

louvers with lips or teeth (Fig. 73). These louvers are mounted on rods and fastened together so that they may be adjusted simultaneously to the desired openings. They are usually available in various shapes of louvers and different spacings to accommodate various crops and conditions.

Non-adjustable chaffers are available in a number of designs. The openings and louvers have different shapes and sizes to meet the problems encountered in different crop conditions. Two types are shown in Fig. 74. In normal conditions, however, the adjustable chaffer will provide the most effective operation.

Chaffer Operation

The grain and chaff mixture is delivered to the front of the chaffer (Fig. 75) over the finger bar or "scalper." The finger bar holds the incoming layer of grain and chaff mixture over the front part of the chaffer and allows the relatively high velocity of air from the fan to break up the layer. The lighter chaff (red) is suspended in air and carried out of the combine. The grain (green) and heavier particles rain down onto the chaffer. The oscillating motion of the chaffer carries those particles and grain toward the rear of the chaffer. The grain and smaller heavy particles fall through the chaffer louvers onto the sieve, and the lighter particles are carried rearward until they either fall through the chaffer extension into the tailings auger or to the ground off the end of the chaffer extension.

Sieve

The sieve is similar to the chaffer except that the louvers and openings are smaller (Fig. 75). The final job of cleaning is done here.

There are several types of sieves, but the two most common types are the *adjustable louver type* (similar

Fig. 74—Two Types of Non-Adjustable Chaffers

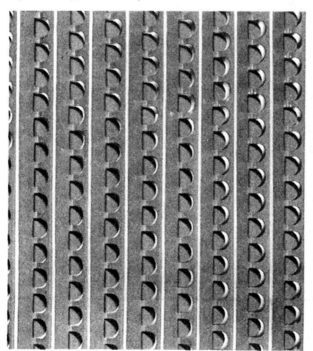

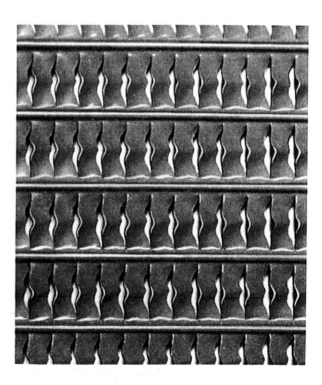

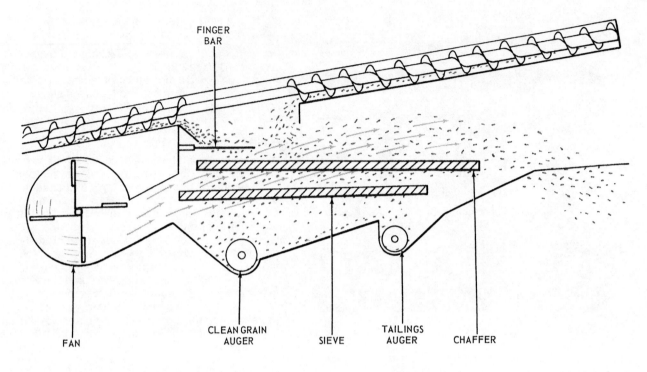

Fig. 75—Chaffer and Sieve Operation

Fig. 76—Handling The Crop

to the chaffer) and the *non-adjustable round-hole type*. The round-hole-type sieve is available in hole sizes from 1/10 inch up to 9/16 inch (2.5 to 14.3 mm) according to the crop being harvested.

The sieve is located below the chaffer and material which falls through the chaffer falls directly onto the sieve (Fig. 75) The sieve oscillates to move this material rearward. In some combines it oscillates with the chaffer and in others it oscillates in a direction counter to the chaffer to help reduce the build-up of straws which poke down through the chaffer. The fan forces air through the sieve to help separate the tailings from the grain. The grain (green) falls through the sieve to the clean grain auger and is carried to the grain tank. The unthreshed grain heads or tailings are carried to the tailings auger by the action of the sieve. The tailings are then carried back to the cylinder for rethreshing.

HANDLING THE CROP

Handling the crop means moving the threshed, separated, and cleaned crop from the cleaning shoe to the grain tank, and then from the grain tank to a wagon or truck for transporting. However, rethreshing of the tailings is another phase of grain handling which must be included.

The crop handling components (Fig. 76) include: lower clean grain auger; clean grain elevator; grain tank loading elevator; lower tailings auger; tailings elevator (not shown); upper tailings auger; grain tank; and grain tank unloading auger.

CLEAN GRAIN ELEVATOR AND AUGERS

After the grain has been cleaned by the shoe, the lower clean grain auger delivers it to the clean grain elevator (Figs. 76 and 77). The elevator carries the grain to the upper clean grain auger or grain tank loading auger which deposits the grain in the center of the grain tank or grain bin.

The augers are usually 4 to 6 inches (10.2 to 15.2 cm) in diameter with the flights 3 to 5 inches (7.6 to 12.7 cm) apart. The elevator has a series of rubber or steel paddles attached to a drive chain which moves at about

Fig. 77—Typical Clean Grain and Tailings Handling System

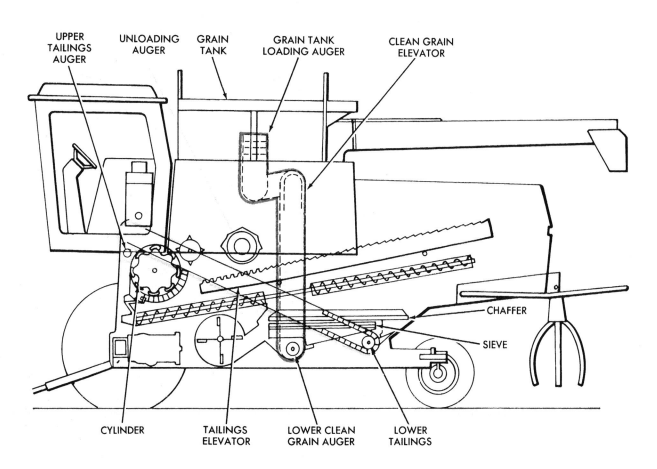

Fig. 78—Grain Tank

350 feet (107 m) per minute. This slow speed helps prevent damage to the grain. In some crops, such as edible beans, metal buckets are used to assure careful handling of the beans.

TAILINGS ELEVATOR AND AUGERS

"Tailings" are unthreshed or unseparated material which fall through the chaffer extension at the end of the chaffer and off the rear of the sieve (Fig. 77). Here the lower tailings auger moves the tailings to the tailings elevator. The elevator then carries the material to the upper tailings auger which drops the tailings in the center of the separator just above the threshing cylinder. Here the material is rethreshed and later separated and cleaned.

The tailings elevator and augers are similar to the clean grain components except that the tailings components are smaller because they do not carry as much material.

Some combines return the tailings to a small tailings-threshing cylinder located near the shoe for re-threshing, and then the rethreshed material is delivered to the chaffer and goes through the cleaning cycle again.

An inspection door is provided so that the operator can examine the tailings to determine whether or not the combine is adjusted correctly. Ideally, there should be very little material returned in the tailings; this indicates a properly adjusted cleaning shoe and that proper threshing is taking place.

GRAIN TANK

The grain tank (Fig. 78) or bin is the storage compartment for the clean grain. Grain tanks come in many shapes and sizes and may be located on top, on one side, or on both sides of the combine. Sizes vary from about 70 to 315 bushels (3 to 11 m³) depending on the size of the combine.

One problem in combine design has been how to increase the capacity of the tank without increasing the height of the combine. Height is a problem because of low storage-building doors and limits on bridges and overpasses when trucks transport combines.

To solve the height problem, engineers have (1) flattened the top of the tank and expanded the sides out; (2) used small tanks (one on each side as a saddle-bag effect); (3) lowered the entire tank; or (4) located the threshing cylinder forward and lower to the ground, allowing more room for the grain tank. All of these variations have greatly reduced the height of the combine and have produced a neat, low-profile machine.

Not only has lowering the grain tank provided a lower profile, but the center of gravity has also been lowered. This has reduced the possibility of tip-over that can occur in taller combines.

To increase capacity of the grain tank, side extensions may be installed. Care must be taken when side extensions are used so that the power train, wheels, and separator are not overloaded. Undue strain may damage these components.

GRAIN TANK UNLOADING AUGER

The auger for unloading should be positioned before beginning to harvest. The auger swing path must be free of obstruction. Keep combine on level ground so auger weight can be handled with minimum effort.

 CAUTION: No one should be in grain tank when the combine engine is running.

When the grain tank becomes full, it is necessary to unload the grain into a wagon or truck (Fig. 79). The harvested grain can then be transported to a storage area or to the elevator for marketing.

An unloading auger system is normally used to unload the grain from the tank. This system usually consists of a large auger across the bottom of the tank (Fig. 78). Connected to it is an outer auger (Fig. 79) which unloads the grain from the tank to the truck. The outer auger is usually tilted at an angle to extend up over the sides of the truck.

On some combines, a sump-type system is used (Fig. 80). Here the grain is augered from the bottom of the tank vertically up the side of the combine. When the grain reaches the top of the vertical auger, it is then carried by a horizontal auger to the truck or wagon.

Because higher yielding crops are being produced today, combine grain tanks now fill much more rapidly and must be unloaded more frequently. Therefore, unloading time must be kept to a minimum. For this reason, large unloading augers (up to 12 inches [30.5 cm] in diameter) are used and are run at a high speed to unload the grain fast. On some combines a 200-bushel (7 m³) tank can be emptied in slightly over a minute and a half.

Fig. 79—Grain Tank Unloading Auger

To save additional time, some combine owners unload "on the go." With this method, the truck or wagon is driven alongside the combine and the grain tank is unloaded as the combine moves through the field (Fig. 81). This allows the combine to keep operating while unloading the tank. On some models, an automatic hydraulic latch allows the operator to swing and latch the unloading auger without leaving the cab to save more time.

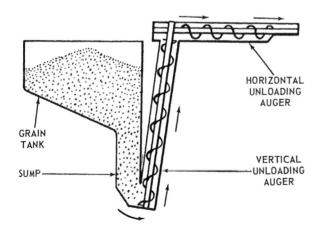

Fig. 80—Sump-Type Unloading Auger

Fig. 81—Unloading Grain Tank "On The Go"

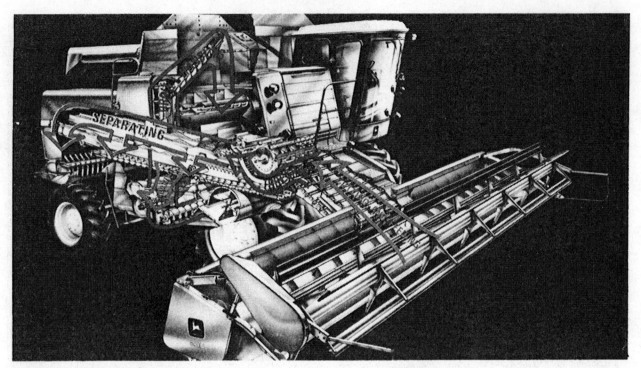

Fig. 82—Flow Of Material Through A Combine

matic hydraulic latch allows the operator to swing and latch the unloading auger without leaving the cab to save more time.

Cooperation between combine operator and hauler is essential. The hauler must position the truck or wagon without getting too close to the combine. The hauler must be prepared for unexpected stops and leave plenty of room for the combine to turn at the ends of the field. The combine operator must stop unloading in time for the hauler to turn corners and drive around obstacles.

SUMMARY OF OPERATION OF COMPONENTS

The operator must know the fundamentals of how a combine works in order to operate it properly. The flow of material through a combine is shown in Fig. 80. In summary, these are the five basic crop harvesting functions of a combine:

1. **Cutting and Feeding**
2. **Threshing**
3. **Separating**
4. **Cleaning**
5. **Handling**

In other chapters, these functions are given more meaning when we discuss field operation, adjustments and troubleshooting.

CHAPTER QUIZ

1. What are the five basic functions of a combine?

2. (Fill in blank.) The mechanism which cuts or gathers the crop and feeds it to the combine separator is known as a _____.

3. (Fill in blank.) Gathering a crop which has been cut and laid into a windrow is accomplished by a _____ platform.

4. (Fill in blank.) A reel and cutter bar are necessary to harvest _____ grains.

5. (Fill in blank.) A series of parallel steel bars held together by curved side bars and mounted under the threshing cylinder is called a _____.

6. What are the two adjustments provided which affect threshing action?

7. True or false? "To reduce the ground speed on a combine, the engine should be slowed down."

8. True or false? "The straw walkers move the grain to the rear of the machine by oscillation."

9. What are the three parts of a cleaning unit on a conventional combine?

10. (Fill in blank.) Grain heads that are not completely threshed at the threshing cylinder are called _____.

3
Power Systems

Fig. 1—Modern Diesel Engine For Self-Propelled Combine

INTRODUCTION

Combines of the past depended on simple forms of power, such as horses, mechanical devices and manpower; today's combines have enormous power requirements. Horses cannot pull a modern combine through the fields efficiently, because the equivalent of 150 work horses would be needed to pull some of the largest machines.

In fact, it takes three power systems to operate a modern self-propelled combine:

- **Engine**
- **Power Train**
- **Hydraulic System**

The ENGINE is the power source; the POWER TRAIN transmits power to the drive wheels or tracks to propel the combine through the field; and the HYDRAULIC SYSTEM converts the engine power to hydraulic force to operate the hydraulic components.

Fig. 2—Modern Combines Contain Diesel Engines

ENGINES

Small self-propelled combines may have engines as small as 70 horsepower (52 kW), while larger machines may require 250 horsepower (185 kW) or more (Fig. 2). Pull-type combines require tractors with about 80 horsepower (60 kW) or more. The engines provide power to propel the machine through the field; to drive the harvesting components and to power the hydraulic systems.

The engine may have four, six or eight cylinders and use either gasoline, LP-Gas or diesel fuel. Engines are mounted in different positions on different combines — on the top, on the side or underneath the combine. Some manufacturers position the engine low to obtain a low center of gravity and provide ground level accessibility. Others prefer to mount it high out of the dust and dirt. This can also provide a better distribution of weight, which is important for good traction.

Fig. 3—Gasoline Engine In Combine

FUEL TYPES

Three types of engines are found on modern combines:

- **Gasoline engine**
- **LP-Gas engine**
- **Diesel engine**

Let's look at the overall performance of each one in a comparison of gasoline and diesel engines.

GASOLINE VS. DIESEL ENGINES

1. *The method of supplying and igniting fuel.*
2. *The higher compression ratio in diesels.*
3. *The generally more rugged design of diesels.*
4. *The grade and type of fuel used.*

Here is a detailed look at each of these differences.

1. Methods of Supplying and Igniting Fuel: Gasoline vs. Diesel

In GASOLINE ENGINES, fuel and air are mixed outside the cylinders in the carburetor or manifold (Fig. 4). The partial vacuum of the piston's intake stroke draws the mixture into the combustion chamber.

In DIESEL ENGINES, there is no premixing of air outside the cylinder. Air only is taken into the cylinder through the intake manifold and compressed (Fig. 4). Fuel is then sprayed into the cylinder and mixed with air as the piston nears the top of its compression stroke.

Gasoline engines use an electric spark to ignite the fuel-air mixture, while in diesels, the heat produced by compression of air results in ignition.

2. Compression Ratios: Gasoline vs. Diesel

Compression ratio compares the volume of air in the cylinder before compression with the volume after compression.

An 8-to-1 compression ratio is typical for gasoline engines, while a 16-to-1 ratio is common for diesels (Fig. 5).

The higher compression ratio of the diesel raises the temperature of the air high enough to ignite the fuel without a spark.

This also gives the diesel better efficiency because the higher compression results in greater expansion of gases in the cylinder following combustion. Result: a more powerful stroke.

The higher efficiency which results from diesel combustion must be offset by the need for sturdier, more expensive parts to withstand the greater forces of combustion.

3. Design of Engine: Gasoline vs. Diesel

We have just touched on the next point: diesels must be built sturdier to withstand the greater forces of combustion. This is generally done by increasing the strength of the pistons, pins, rods, and cranks, and by more or larger main bearings to support the crankshaft.

4. Grades and Types of Fuel: Gasoline vs. Diesel

Fuel energy is measured in standard heat units or "British Thermal Units" (BTU) (or kJ) and gives a comparison of the power possible from each fuel.

Diesel fuel has more heat units (BTU) per gallon (kJ/L) and so gives more work per gallon (liter) of fuel. In addition, diesel fuel normally costs less than gasoline.

However, diesel fuel-injection equipment is more expensive than gasoline equipment.

In selection of a fuel type for the engine, a deciding factor often is how much fuel is consumed per year in the engine operation. The more fuel that is burned, the greater the potential savings with the greater efficiency of the diesel engine.

LP-GAS ENGINES

The LP-gas engine is similar to the gasoline model, but requires special fuel handling and equipment.

LP-gas engines have higher compression ratios than gasoline engines but not as high as diesels.

In areas where LP-gas fuel is available at low prices, these engines are very popular. However, in many areas, LP-gas fuel is not competitive in price with the other fuels.

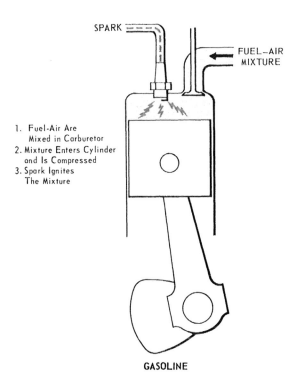
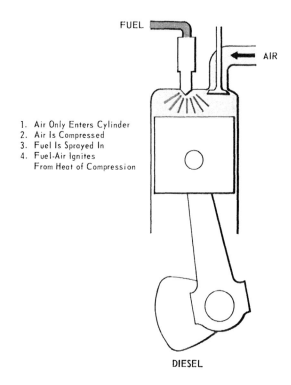

Fig. 4—Methods of Supplying and Igniting Fuel

SUMMARY: COMPARING ENGINES

The chart below compares gasoline, LP-gas, and diesel engines.

The comparisons assume that each fuel is available at reasonable prices. Performance is based on general applications which are suited to the engine and fuel type. It is also assumed that the engines are all in good condition.

COMPARING THE ENGINES

	Gasoline	LP-Gas	Diesel
Fuel Economy	Fair	Good	Best
Hours Before Maintenance	Fair	Good	Good
Weight per Horsepower (kW)	Low	Low	High
Cold-Weather Starting	Good	Fair	Fair
Acceleration	Good	Good	Fair
Continuous Duty	Fair	Fair	Best
Lubricating-Oil Contamination	Moderate	Lowest	Low

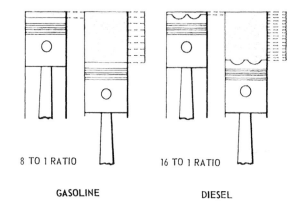

Fig. 5—Engine Compression Ratios Compared

OPERATION OF ENGINES

Here is a brief description of how an engine operates. For more information, refer to John Deere's *Fundamentals of Service Manual — Engines.*

For an engine to operate, a definite series of events must occur in the correct sequence. They are:

1. *Fill the cylinder with a combustible mixture.*
2. *Compress this mixture into a smaller space.*
3. *Ignite the mixture and cause it to expand, producing power.*

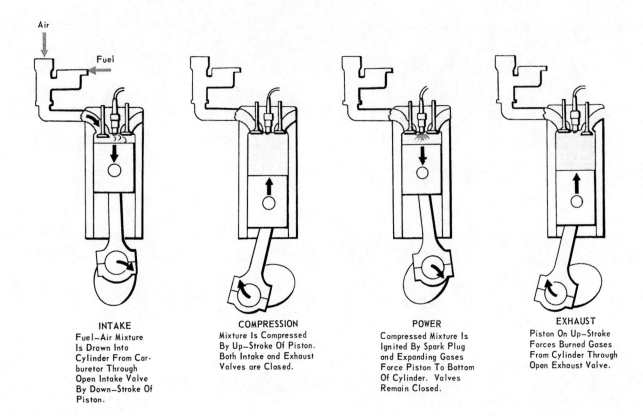

Fig. 6—Four-Stroke Cycle Engine (Gasoline Shown)

4. *Remove the burned gases from the cylinder.*

The sequence above is generally called:

- **Intake**
- **Compression**
- **Power**
- **Exhaust**

To produce sustained power, the engine must repeat this sequence over and over again. One complete series of these events in an engine is called a *cycle.*

Most engines on modern combines are *four-stroke-cycle* engines. Four strokes of the piston—two up and two down—are needed to complete the cycle. As a result, the crankshaft will rotate *two* complete turns while *one* cycle is being completed (Fig. 6).

INTAKE STROKE

The intake stroke starts with the piston near the top and ends as it reaches the bottom of the cylinder. The intake valve is opened, allowing the cylinder, as the piston moves down, to receive the fuel-air mixture. The valve is then closed, sealing the cylinder.

COMPRESSION STROKE

The compression stroke begins with the piston at bottom and rising up to compress the fuel-air mixture. Because the intake and exhaust valves are closed, there is no escape for the fuel-air and it is compressed to a fraction of its original volume.

POWER STROKE

The power stroke begins as the piston is almost at the top of the cylinder in the compression stroke and the fuel-air mixture is ignited. As the mixture burns and expands, it forces the piston down on its power stroke. The valves remain closed so that all the force is exerted on the piston.

EXHAUST STROKE

The exhaust stroke begins when the piston nears

the end of its power stroke. The exhaust valve is opened and the piston rises, pushing out the burned gases. When the piston reaches the top, the exhaust valve closes and the piston begins a new four-stroke cycle of intake, compression, power, and exhaust.

As it completes the cycle, the crankshaft has gone all the way around *twice*.

ENGINE SYSTEMS

Here are other systems required for engine operation:

- **Fuel Systems**
- **Intake and Exhaust Systems**
- **Lubricating Systems**
- **Cooling Systems**
- **Electrical Systems**
- **Governing Systems**

We will now discuss each of these systems.

ENGINE FUEL SYSTEMS

A fuel system must deliver clean fuel, in the quantity required, to the fuel intake of an engine. It must also provide for safe fuel storage and transfer.

The three combine fuel systems are:

- **Gasoline**
- **LP-Gas**
- **Diesel**

GASOLINE FUEL SYSTEM

The gasoline fuel system supplies a combustible mixture of fuel and air for the engine.

The basic gasoline fuel system (Fig. 7) has three parts.

- **Fuel Tank—stores fuel**
- **Fuel Pump—moves fuel to carburetor**
- **Carburetor—atomizes and mixes fuel with air**

In operation, the FUEL PUMP moves gasoline from the tank to the carburetor bowl.

The CARBURETOR is basically an air tube which atomizes fuel and mixes it with air by a difference in air pressure. It meters both the fuel and air for the engine.

On its intake stroke, the engine creates a partial vacuum. This allows outside air pressure to force the fuel-air vapor mixed in the carburetor into the engine cylinder for compression and combustion.

Fig. 7—Gasoline Fuel System

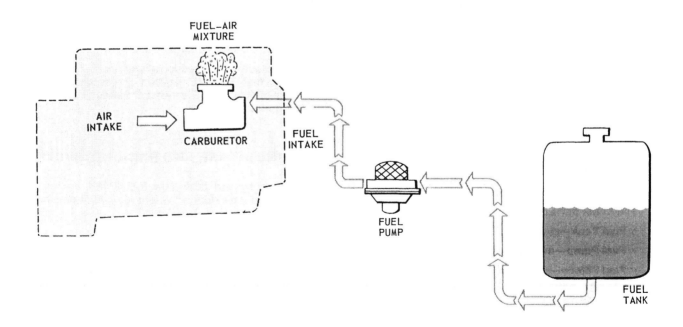

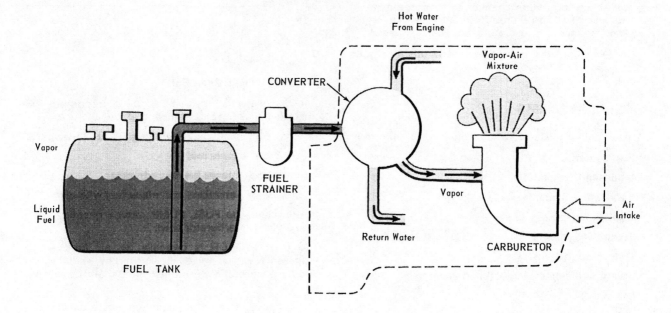

Fig. 8—LP-Gas Fuel System

LP-GAS FUEL SYSTEM

The LP-gas fuel system (Fig. 8) also supplies a combustible mixture of fuel and air to its engine. However, LP-gas vaporizes at low temperatures. Thus the fuel tank must be a closed unit to prevent vapor from escaping.

The LP-gas carburetor is simpler than the gasoline type, since the fuel is already vaporized. It meters the vapor and mixes it with the proper amount of air for the engine.

DIESEL FUEL SYSTEM

In the diesel fuel system (Fig. 9), fuel is sprayed directly into the engine combustion chamber where it mixes with hot compressed air and ignites. No electrical spark is used to ignite the mixtures (as in gasoline and LP-gas engines).

Instead of a carburetor, a fuel injection pump and injection nozzles are used.

The major parts of the diesel fuel system are:

- **Fuel Tank—stores fuel**
- **Fuel Pump—moves fuel to injection pump**
- **Fuel Filters—help clean the fuel**
- **Injection Pump—times, measures, and delivers fuel under pressure**
- **Injection Nozzles—atomize and spray fuel into cylinders**

In operation, the FUEL PUMP moves fuel from the tank and pushes it through the FILTERS. Clean fuel free of water is very vital to the precision parts of the diesel injection system. Extra filters are often used to assure clean fuel, but buying clean fuel and storing it properly are also prime needs for efficient operation.

The fuel is then pushed on to the INJECTION PUMP where it is metered, put under high pressure, and delivered to each INJECTION NOZZLE in turn.

The nozzles each serve one cylinder; they atomize the fuel and spray it under controlled high pressure into the combustion chamber at the proper moment.

High-pressure fuel is needed at each nozzle to get a fine spray of fuel. This atomization of fuel assures good mixing of fuel with the hot compressed air for full combustion.

ENGINE INTAKE AND EXHAUST SYSTEMS

The intake system carries the fuel-air mixture into the engine and the exhaust system removes the exhaust gases after combustion (Fig. 10).

INTAKE SYSTEM

The intake system supplies the engine with clean air of the proper quantity, temperature, and mix for good combustion.

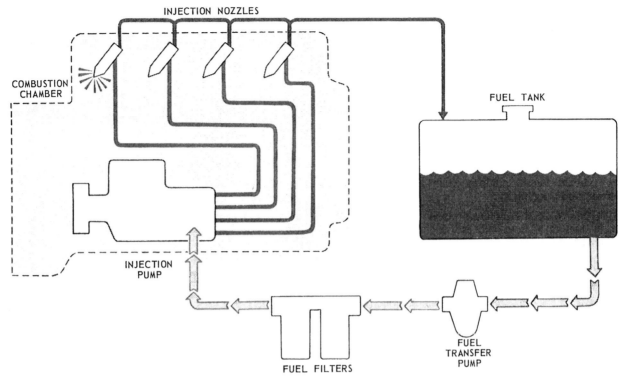

Fig. 9—Diesel Fuel System

Fig. 10—Intake and Exhaust Systems

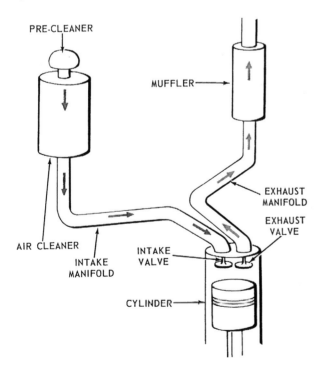

Fig. 11—Engine Lubricating System

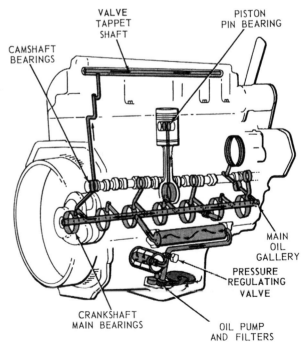

The intake system has four to six parts:

- **Air cleaners**
- **Turbocharger (if used)**
- **Carburetor air inlet**
- **Intake manifold**
- **Intake valves**
- **Intercooler (if used)**

AIR CLEANERS filter dust and dirt from the air passing through them enroute to the carburetor. Precleaners prevent larger particles from reaching the air cleaner and plugging it.

TURBOCHARGERS increase power by forcing more air or fuel-air mixture into the engine cylinders. See the turbocharger system shown in Fig. 1.

AIR INLET supplies fuel mixed with incoming air in the proper ratio for combustion, and also controls engine speed. On spark-ignition engines, this mixture comes from the carburetor. On diesel engines, air only is provided, with fuel injected into the engine cylinders later.

INTAKE MANIFOLDS transport the fuel-air mixture (pure air on diesel engines) to the engine cylinders.

INTAKE VALVES admit air to diesel engines and the fuel-air mixture to spark-ignition engines. They are normally opened and closed by mechanical linkage from the camshaft.

INTERCOOLERS help increase power, improve fuel efficiency and reduce engine noise by using engine coolant to reduce the temperature and increase the density of air compressed by the turbocharger.

EXHAUST SYSTEMS

The exhaust system collects the exhaust gases after combustion and carries them away.

An exhaust system has three basic parts:

- **Exhaust valves**
- **Exhaust manifold**
- **Muffler**

EXHAUST VALVES open to release the burned gases on four-cycle engines. The valves are normally driven by the camshaft.

The EXHAUST MANIFOLD collects the exhaust gases and conducts them away from the cylinder.

Fig. 12—Engine Cooling System

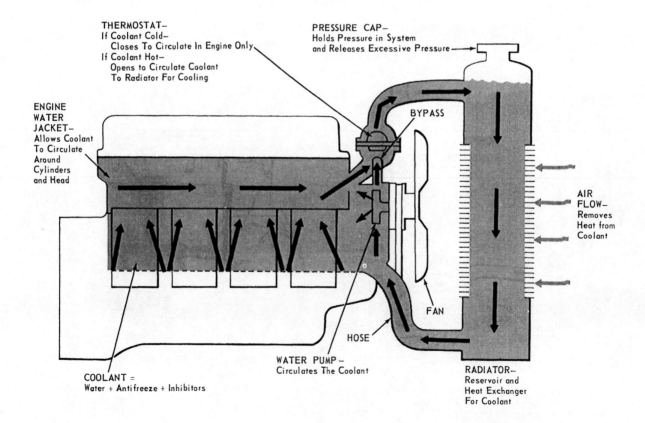

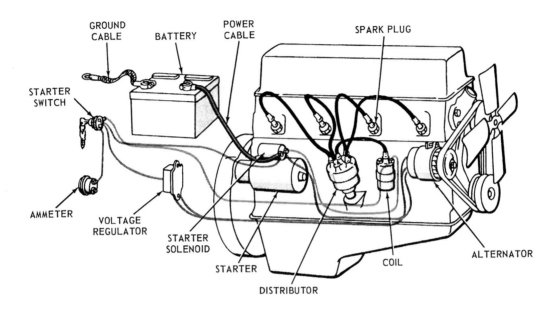

Fig. 13—Engine Electrical Systems (Spark-Ignition Engine)

The MUFFLER reduces the emission sounds of the engine during the exhaust cycle.

ENGINE LUBRICATING SYSTEM

The engine lubricating system (Fig. 11) reduces friction, dissipates engine heat, and helps keep the engine parts clean.

The lubricating system has these parts:

- **Crankcase oil reservoir**
- **Oil pump**
- **Oil filter**
- **Oil passages**
- **Pressure regulating valve**

The CRANKCASE OIL RESERVOIR holds the oil which circulates in the engine system. Oil is pushed through the system by the OIL PUMP, and oil is cleaned by the OIL FILTER, which must be either cleaned or replaced periodically. Oil reaches engine parts through OIL PASSAGES as shown in Fig. 11. Pressure of oil is controlled by the PRESSURE REGULATING VALVE.

ENGINE COOLING SYSTEM

The cooling system prevents overheating of the engine and regulates its temperature at the best levels.

The cooling system (Fig. 12) normally uses water as a coolant. In cold weather, anti-freeze solutions are added to the water to prevent freezing. The coolant circulates in a jacket around the cylinders and through the cylinder head. As heat radiates, it is absorbed by the coolant, which then flows to the radiator. Air flow through the radiator cools the coolant and dissipates heat into the air. The coolant then recirculates into the engine to pick up more heat.

ENGINE ELECTRICAL SYSTEMS

The design of the electrical systems for a combine is similar to that for other types of farm equipment (Fig. 13).

The complete engine electrical system consists of three circuits:

- **Charging Circuit**
- **Starting Circuit**
- **Ignition Circuit (Spark-Ignition Engines)**

Here is a brief look at these circuits.

CHARGING CIRCUIT

The charging circuit recharges the battery and generates current during engine operation. It has these parts:

- **Battery**
- **Voltage Regulator**
- **Alternator (or Generator)**

D.C. charging circuits have a *generator* and a *regu-*

lator. The generator supplies the electrical power and rectifies its current mechanically by using a commutator and brushes.

The regulator has three functions: (1) opens and closes the charging circuit; (2) prevents overcharging of the battery; (3) limits the generator's output to safe rates so as not to damage the electrical system.

A.C. charging circuits have an *alternator* and a *regulator*.

The ALTERNATOR is really an a.c. generator. Like the generator, it produces alternating current but rectifies it electronically using diodes. Alternators are generally more compact than generators of equal output, and supply a higher current output at low engine speeds. They have replaced generators on most modern engines.

The REGULATOR in A.C. charging circuits limits the alternator voltage to a safe, preset level. Transistorized models are used in many of the modern circuits.

STARTING CIRCUIT

The starting circuit converts electrical energy from the battery into mechanical energy at the starting motor to crank the engine.

A basic starting circuit has four parts:

1. The *BATTERY* supplies energy for the circuit.
2. The *STARTER SWITCH* actuates the circuit.
3. The *MOTOR SWITCH* (solenoid) engages the motor drive with the engine flywheel.
4. The *STARTING MOTOR* drives the flywheel to crank the engine.

IGNITION CIRCUIT (Spark-Ignition Engines)

The ignition circuit creates the spark which ignites the fuel-air mixture that powers the gasoline or LP-gas engine.

The ignition circuit has these parts:

- **Ignition Coil**
- **Condenser**
- **Breaker Points**
- **Distributor**

Fig. 14—Typical Electrical System On Modern Combine

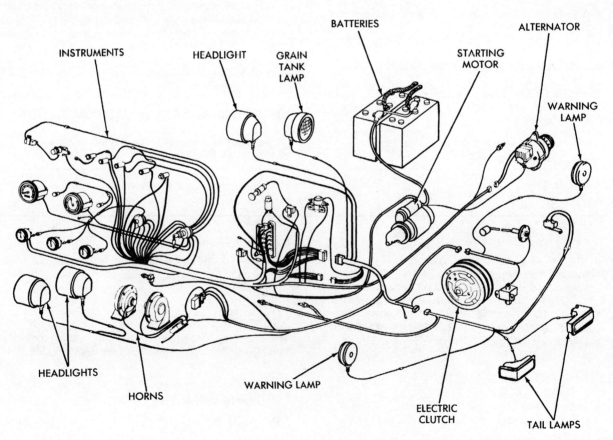

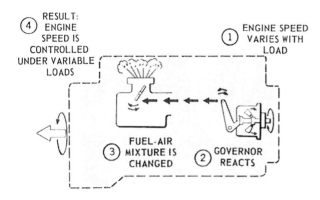

Fig. 15—Engine Governing System (Gasoline Engine Shown)

- **Spark Plugs**
- **Ignition Switch**

The COIL transforms the low voltage from the battery to a high voltage for producing a spark.

The CONDENSER aids in collapsing the magnetic field in the coil to produce a high voltage. In doing this it also protects the distributor points against arcing by absorbing the current surge in the primary circuit.

The BREAKER POINTS open and close the primary circuit, causing the coil to produce high voltage surges. The DISTRIBUTOR times these surges to engine rotation, and directs each high-voltage surge to the proper spark plug.

The SPARK PLUGS ignite the fuel-air mixture within each cylinder of the engine.

The IGNITION SWITCH actuates the ignition circuit. The battery is the initial power source for the voltage in the ignition circuit, while the ignition switch turns on the circuit when it cranks the engine.

ELECTRICAL POWER FOR ACCESSORIES

The battery which supplies power for the engine also supplies power for the accessories of a modern combine. Many years ago, on engine-driven combines, the only electricity required was to start and run the engine. Now modern combines require much more electrical power to operate lights, air conditioning, radios, horns, instruments, and electrical clutches.

Complex electrical systems are required to operate today's combines. A typical wiring diagram for a self-propelled combine is shown in Fig. 14. Electrical power for pull-type combines is supplied by the tractor's electrical system.

ENGINE GOVERNING SYSTEMS

The governing system (Fig. 15) keeps the engine speed at a constant level. It does this by varying the amount of fuel or fuel-air mixture supplied to the engine, according to the demands of the load. The level of engine speed is controlled by the position of the speed control lever, connected by linkage to the governor. On diesel engines, the governor is built into the fuel injection pump.

The function of the governor is to get the engine's power to match the load at all times, to keep the speed at a steady level.

Governors can be either mechanical, hydraulic, electrical, or electronic.

TRANSMITTING ENGINE POWER

Most combines have a main or *primary countershaft*

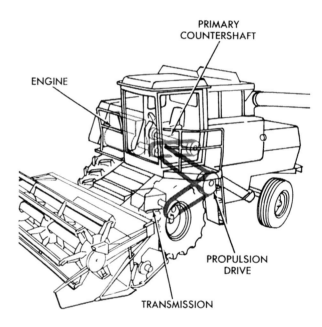

Fig. 16—Primary Countershaft Drive

(driven by the engine) from which the various units of the combine are driven (Fig. 16). This shaft is the prime shaft for operating the components of the combine at the proper speed. The engine must operate at the speed specified by the manufacturer to make this shaft run at the proper speed. (For pull-type combines, the tractor PTO must be operated at the proper speed.) Before making any adjustments, the countershaft speed must be checked. If the primary countershaft is not run at the designated speed, the components of the combine will not function efficiently. This may cause poor cutting, feeding, threshing, separating or cleaning. All of these will affect the capacity of the combine.

Most combines have three basic drive systems:

- **Propulsion Drives**
- **Header Drives**
- **Separator Drives**

Each system may have one or more drives to provide power for its components.

PROPULSION DRIVES

The propulsion drive transmits power from the engine to the power train to propel the combine. On combines which have belt-driven power trains, a *variable-diameter sheave* arrangement is generally used to provide speed ranges in each gear (Fig. 17). From the engine powershaft (a shaft connected directly to the engine crankshaft), one belt drives one side of the variable sheave assembly. Another belt on the other side of the variable-sheave assembly drives the input shaft of the transmission. The clutch is attached to this input shaft.

The operator controls the combine movement by disengaging or engaging the transmission clutch and shifting the transmission into the desired gear. After the combine is in motion, the operator can hydraulically control the variable-speed sheave assembly to give either a low- or high-range speed in the gear selected (Fig. 17).

Some combine belt drives (propulsion, cylinder, and other components) have automatic tensioning systems which adjust belt tension to match the load. This extends belt life by reducing stretch and wear. It also saves power by reducing slippage as well as shaft and bearing load caused by constant tension high enough to prevent slippage under all operating conditions.

If the combine is hydraulically propelled, a similar drive may be used except the variable sheaves are

Fig. 17—Variable Sheave Operation

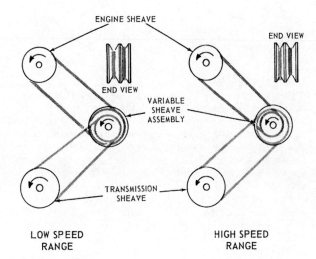

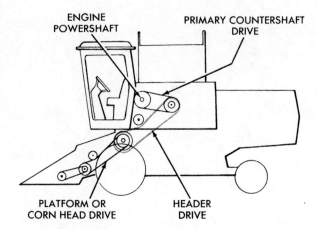

Fig. 18—Header Drive System On Typical Self-Propelled Combine

not required. The drive belts simply connect the engine powershaft with the hydraulic drive pump. Other hydraulic drive combines do not use belts to drive the pump; the pump is driven directly from the engine powershaft.

A track-type drive may be used instead of wheels for very soft or muddy fields, particularly when harvesting rice. On some combines, track and wheel drives are interchangeable, but other models must be ordered from the factory with one or the other type of drive. Power flow to either tracks or wheels is essentially the same.

Later in this chapter, power trains will be discussed in more detail.

ENGINE GEARCASE OPERATION

The engine gearcase steering pump (Fig. 19) draws pressure-free oil from the gearcase sump. Regulated pressure is maintained by a regulator valve. Regulated pressure oil flows from the pump to the steering control valve. Regulated pressure oil flows from the steering system, through the engine gearcase filter and into the control valve. Oil from the regulator valve flows through the oil cooler.

Lubrication pressure is maintained by a lube pressure valve. The filter bypass valve allows continued lube oil flow if the filter is plugged. An oil cooler bypass valve allows oil flow to continue to the lube circuit, if oil flow is restricted through the cooler.

The unloading auger engage cylinder has pressure free oil when disengaged. When engaged, regulated pressure oil is sent to the unloading auger engage cylinder.

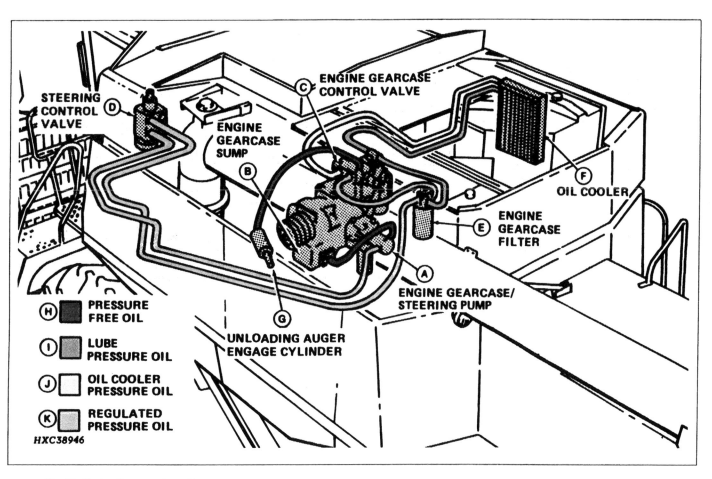

Fig. 19—Engine Gearcase Operation

HEADER DRIVES

The header is driven by the primary countershaft (Fig. 18). The operator engages or disengages the header drive either by a mechanical control or a switch which controls an electrical clutch. The drive mechanism may be either belt drives, chain drives, or a combination of both. To illustrate the principles of basic header drive systems, we will describe simplified drive systems. Keep in mind that any of these drives may be either belt or chain drives, or a combination of both.

Feeder Conveyor and Header Drive

The overall header drive system is shown in Fig. 18, while a typical feeder conveyor and header drive is shown in Fig. 20. The feeder conveyor is driven by the drive from the main powershaft. In turn, the header drive sheave is mounted on the feeder conveyor shaft and transmits power to the header drive shaft.

In some conditions, the speed of the header drive must be changed to gather the crop at different rates, depending on crop condition and forward speed of the combine. Some manufacturers provide different sizes of sheaves to change speeds; others make variable-diameter sheaves available.

The header drive shaft is coupled to the drive shaft on the platform or corn head, on machines where the platform or corn head is detachable from the feeder conveyor unit (Fig. 19). On machines where the platform or corn head is permanently attached to the feeder conveyor unit, the shaft is usually one piece.

Cutting Platform Drive

As mentioned above, the platform is driven by the header drive shaft (Fig. 20). At the outer end of the header drive shaft, a combination sprocket-crank may be used to drive the knife and auger (Fig. 21).

The knife drive consists of a *pitman* which is connected to the crank located at the rear of the platform and a *bellcrank* located near the front of the platform. As the drive shaft rotates, the crank moves the pitman back and forth at several hundred strokes per minute, which causes the bellcrank to move the knife back and forth rapidly to cut the crop.

The auger drive is usually powered by a chain from the combination sprocket-crank. Different sizes of sprockets are usually available to vary the speed of the auger.

The reel drive (Fig. 21) usually consists of two drives: (1) a chain drive from the auger to the reel drive countershaft, and (2) a belt drive from the countershaft to the reel shaft (pickup reels may have either a belt or a chain drive). The reel speed on some combines is adjustable by changing the size of sprocket or sheave on the reel drive countershaft.

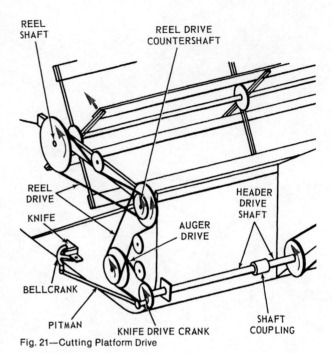

Fig. 21—Cutting Platform Drive

Some manufacturers offer variable diameter sheaves or hydraulic motors to drive and control the reel speed. In these two cases, the reel speed is usually controlled from the operator's station.

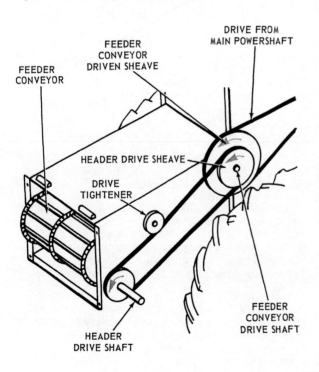

Fig 20—Feeder Conveyor And Header Drive

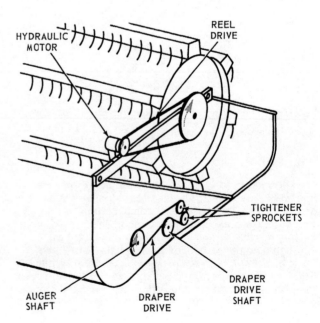

Fig. 22—Draper Drive And Pickup Reel Drive

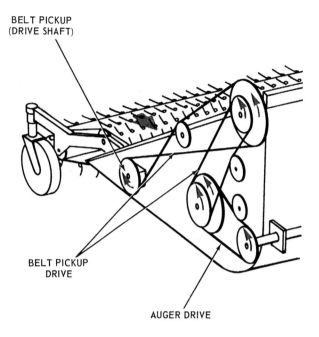

Fig. 23—Pickup Drive System

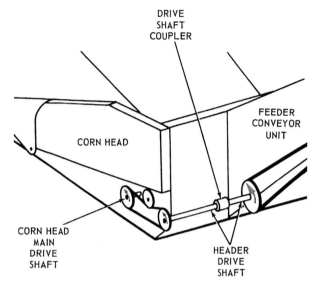

Fig. 24—Corn Head Drive

The draper drive on rice platforms is usually driven from the auger shaft on the other end of the platform, opposite the knife and auger drive (Fig. 22). The draper is usually driven by a chain. Notice the direction of rotation on the draper drive shaft sprocket.

The reel drive in Fig. 22 consists of a hydraulic motor and chain drive. Another motor drives the reel from the other side of the platform; the motors are connected in series so that each is turning the reel at the same speed as the other.

Pickup Platform Drive

Basically, the drive for the pickup platform is the same as for the cutting platform, except the drive that normally delivers power to the reel is also used to run the belt pickup. Notice in Fig 23 that the belt pickup turns opposite to the direction of the reel.

Corn Head Drive

The corn head is driven by the header drive shaft (see Figs. 20 and 24). At the outer end of the header drive shaft is a chain drive which delivers power to the corn head main drive shaft (Fig. 24). This main drive shaft drives the gear units for each row of the corn head. Both the gathering chains and stalk rolls are driven in turn by the gear units.

SEPARATOR DRIVES

All harvesting components of the combine are driven from the primary countershaft or the secondary countershaft, which is driven by the primary countershaft. See Figs. 25 and 26. On self-propelled combines, all the separator drives (except the unloading auger drive) are controlled by one control lever which engages or disengages the primary countershaft drive. On pull-type machines, the separator drives are controlled by engaging or disengaging the PTO.

Here are the harvesting drives listed by function:

Threshing Drive:
- *Threshing Cylinder*

Separating Drives:
- *Beater*
- *Straw Walkers*

Cleaning Drives:
- *Shoe Supply Conveyor*
- *Shoe*
- *Fan*

Grain Handling Drives:
- *Clean Grain Elevator and Augers*
- *Tailings Elevator and Augers*
- *Unloading Auger*

Straw Disposal Drives:
- *Straw Chopper*
- *Straw Spreader*

The drives shown in Figs. 25, 26, 27 and 28 are simplified for clarity. Most drives are more complex than those shown; for example, drive tighteners are not shown and drive configurations are simplified.

Study Figs. 25 and 26 to learn the operation of these drives.

Threshing Cylinder Drive

The threshing cylinder is usually driven by the primary countershaft (Fig. 25). In order to work the combine in varying crop conditions, the operator must be able to adjust the threshing cylinder speed. One method is to vary the effective diameter of the sheave that drives the cylinder. A split sheave half can be pulled toward or pushed away from a fixed half. Sheaves that are closed cause the drive belt to ride higher, create more speed, and less torque on a smaller threshing cylinder sheave (Fig. 27).

A second way to control the threshing cylinder speed is to use a gear assembly on the cylinder shaft. In this example, the cylinder speed and torque depends upon the gear ratios that are chosen. Both the split sheave and gear box methods can be used simultaneously.

Beater Drive

The beater is usually driven from the cylinder shaft so that the beater rotates either faster or slower as the speed of the cylinder is increased or decreased (Fig. 25). Thus, the speed of the beater is controlled by the speed of the cylinder.

Straw Walker Drive

On some combines the straw walkers are driven from the secondary countershaft (Fig. 26). The speed of the walkers remains constant, because the secondary countershaft is driven at a fixed speed by the primary countershaft. Notice that the drive belt for the straw walkers is twisted to turn the walker cranks in the opposite direction of most shafts; this is necessary so that the walkers can move material out the back of the combine. Straw walkers are also driven through a gearbox. This eliminates sprockets and chains.

Cleaning Shoe Supply Conveyor Drives

In the example shown, augers are driven by a belt drive from the secondary countershaft (Fig. 25). The same basic principle is used with a conveyor chain drive.

Shoe Drive

The shoe, in this instance, is driven off the secondary countershaft (Fig. 26). A crank is usually driven by a chain from the countershaft. Connected to the crank is a pitman, which imparts the shaking motion to the shoe (see Figs. 26 and 28).

Fig. 25—Drives on Right-Hand Side Of Typical Combine

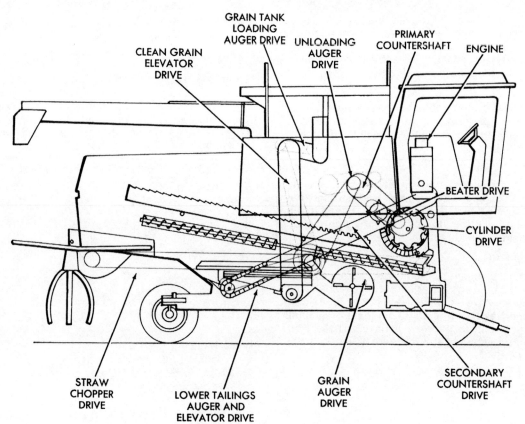

Fan Drive

The cleaning fan is driven from the secondary countershaft also. Most combines are equipped with a variable-speed fan drive which provides infinite speed control to clean various crops under different conditions (Fig. 28). The variable sheaves are similar to those used on the cylinder drive.

Clean Grain Elevator and Augers

The clean grain elevator on a typical combine is driven from the secondary countershaft (see Fig. 25). Here the elevator is driven from the top, while the conveyor chain inside the elevator drives the lower auger. The upper auger is driven by a separate drive from the top elevator shaft.

Tailings Elevator and Augers

The tailings elevator and augers are also driven from the secondary countershaft (Fig. 25). The elevator is driven through the lower tailings auger, while the upper tailings auger is driven by the elevator conveyor chain.

Unloading Auger Drive

The unloading auger countershaft is driven from the engine powershaft (see Fig. 26); the unloading auger countershaft goes through the combine to the other side where another drive connects it to the unloading auger (see Fig. 25). The operator engages or disengages the unloading auger drive with a lever or by actuating an electric clutch. This permits unloading the tank without operating the entire machine.

Straw Chopper Drive

The straw chopper is also driven from the secondary countershaft (Fig. 25).

Straw Spreader Drive

The straw spreader is driven from the rear straw walker crank (Fig. 26).

POWER TRAINS

On self-propelled combines, engine power is transmitted to the drive wheels or tracks by the power train (Fig. 29). It does four basic jobs:

1) *Connects and disconnects power*
2) *Selects speed ratios*
3) *Provides a means of reversing*
4) *Equalizes power to drive wheels for turning*

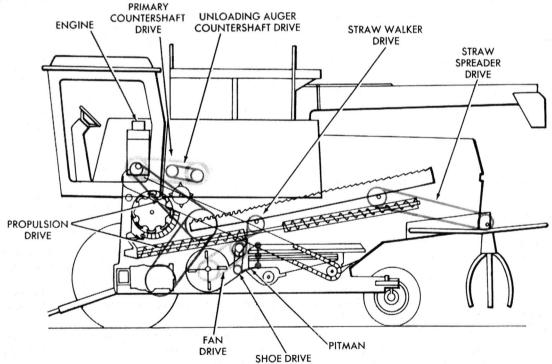

Fig. 26—Drives on Left-Hand Side Of Typical Combine

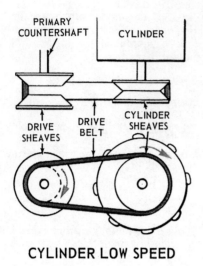
CYLINDER LOW SPEED

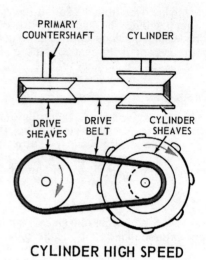
CYLINDER HIGH SPEED

Fig. 27—Cylinder Drive Sheaves

To perform these functions, five basic parts (see Fig. 30) are needed:

- **Clutch — to connect and disconnect power**
- **Transmission — to select speeds and direction**
- **Differential — to equalize power for turning**
- **Final Drives — to reduce speed and increase torque to axles**
- **Drive Wheels or Tracks — to propel the machine**

We will now discuss these parts in more detail.

CLUTCH

The clutch is a vital link in the power train. It connects and disconnects power between the engine and the transmission.

A clutch is used in all combines except those which have hydraulic drives.

Most combines use the standard dry-type clutch with mechanical linkage shown in Fig. 31.

In the *engaged* position, the pressure plate provides pressure against the clutch plate and forces the plate against the flywheel. Power is transmitted from the engine through the flywheel and clutch plate to the drive shaft or transmission input shaft.

The clutch is *disengaged* by applying pressure on the pedal which pushes the clutch release bearing assembly against the release levers. The levers pull the pressure plate away from the clutch plate so that the

Fig. 28—Typical Variable Speed Fan Drive

clutch plate is no longer forced against the flywheel. At this point the flywheel and pressure plate are free to rotate independently of the clutch plate and drive shaft and power flow from the engine is disengaged.

With *hydraulically operated clutches,* a master cylinder similar to a brake master cylinder is attached to the clutch pedal. A slave cylinder is connected to the master cylinder by a flexible pressure hose or metal tubing. The slave cylinder is also connected to the clutch release mechanism.

Movement of the clutch pedal actuates the clutch master cylinder. This movement is transferred by hydraulic pressure to the slave cylinder, which in turn actuates the clutch release mechanism, disengaging the clutch.

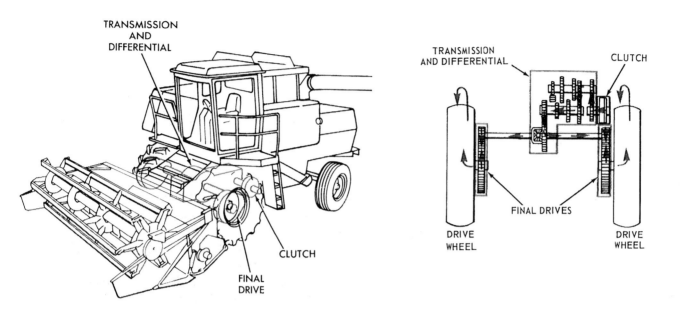

Fig. 29—Power Train Of A Typical Combine

Fig. 30—Flow Of Power Through The Power Train

Fig. 31—Engine Clutch In Operation

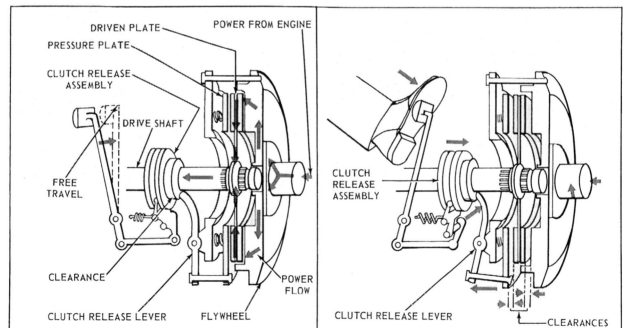

TRANSMISSION

The transmission has a train of gears that transfers and adapts the engine power to the drive wheels of the combine.

The transmission has two functions:

- **Selects speed ratios for various travel speeds**
- **Reverses travel of the combine**

A typical combine transmission allows ground speeds from less than one mph (0.6 km/h) to nearly 20 mph (32 km/h). Refer to the chart below for speed ranges in each gear for a typical combine.

TYPICAL GROUND SPEEDS			
	1st Gear	2nd Gear	3rd Gear
MPH	.8 to 2.0	1.7 to 4.5	3.0 to 9.0
(km/h)	(1.3 to 3.2)	(2.7 to 7.2)	(4.8 to 14.5)
	4th Gear	Reverse Gear	
MPH	7.5 to 19.0	1.5 to 3.5	
(km/h)	(12.1 to 30.6)	(2.4 to 5.6)	

Normally the transmission is mounted on the rear of the drive axle (Fig. 32). The differential is housed in the same housing as the transmission gears, and the clutch housing is attached. The transmission may have three or four forward gear speeds.

Combines are usually equipped with either of these types of transmissions:

- **Sliding Gear**

Fig. 32—Transmission On Typical Combine

- **Collar Shift**

The SLIDING GEAR transmission has two or more shafts mounted in parallel or in line, with sliding spur gears arranged to mesh with each other and provide a change in speed or direction.

The COLLAR SHIFT transmission (Fig. 33) has parallel shafts with gears in constant mesh. Shifting is done by using sliding collars to lock free-running gears to their shafts. Gears with helical gear teeth are used, which gives quieter operation.

DIFFERENTIAL

The differential has two functions:

- **Transmits power "around the corner" to the drive axles.**
- **Allows each drive wheel to rotate at a different speed and still propel its own load.**

The ring gear and bevel gears direct the power to the axles, while the bevel pinions give the differential action (Fig. 34).

Fig. 33—Collar Shift Transmission In Combine

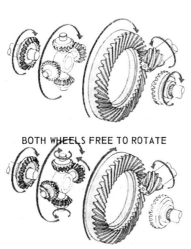

BOTH WHEELS FREE TO ROTATE

ONE WHEEL FREE TO ROTATE

Fig. 34—Differential in Operation

When the combine is moving straight ahead, *both wheels are free to rotate* as shown in Fig. 34.

Engine power comes in on the pinion gear and rotates the ring gear. The four bevel pinions and the two bevel gears are carried around by the ring gear and all gears rotate as one unit. Each axle receives the same rotation and so each wheel turns at the same speed.

When the machine turns a sharp corner, only *one wheel is free to rotate* as shown in Fig. 34.

Again engine power comes in on the pinion gear and rotates the ring gear, carrying the bevel pinions around with it. However, the right-hand axle is held stationary and so the bevel pinions are forced to rotate on their own axis and "walk around" the right-hand bevel gear.

FINAL DRIVES

Final drives, as the last components of the power train, give the final reduction in speed and increase in torque to the drive wheels.

On most combines, the final drives are mounted near the drive wheels (Fig. 35) to avoid the stress of long axle shafts.

By reducing speeds, the final drives lower the stress and simplify the transmission, since extra gears and shafts can be eliminated.

Most final drives must support the weight of the machine as well as withstanding torque and shock loads.

There are two types of final drive axles:

- **Rigid Axle Shaft**
- **Flexible Axle Shaft**

The RIGID AXLE SHAFT (Fig. 36) is connected to the differential output by a splined coupling. Level-land combines use this type.

The FLEXIBLE AXLE SHAFT (Fig. 36) is used when the drive wheels are independently suspended. The axles are connected to the differential by universal joints. This type of axle is sometimes referred to as a "swing axle." The drive wheel is free to move vertically without affecting the position of the differential and transmission. Flexible axles are commonly found on hillside combines.

Fig. 35—Final Drive

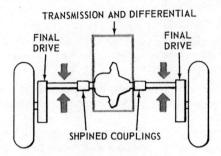

RIGID-AXLE SHAFT
(Level-Land Combine)

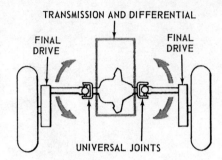

FLEXIBLE AXLE SHAFT
(Hillside Combine)

Fig. 36—Final Drive Axle Construction

HYDROSTATIC DRIVES

Combines may be equipped with hydraulic or "hydrostatic" drives rather than conventional transmissions to propel the machine (Fig. 37). The hydrostatic drive is an automatic fluid drive which uses fluid under pressure to transmit engine power to the drive wheels of the combine (Fig. 38).

Hydrostatic drive combines are used in crop or field conditions which require considerable stop-and-go driving or downshifting, such as when combining rice. There is no clutch, and one lever provides infinite speeds from zero to top speed in each gear, whether in forward or reverse. Also, the speed remains constant as the combine travels either uphill or downhill.

To operate a hydrostatic drive combine, the operator places the speed range control lever in neutral and shifts the transmission into the desired gear. Then he only has to move the speed control to move the combine. As he moves the lever forward, the combine travels forward — the farther he moves the lever, the faster the combine travels. In reverse the opposite is true, except the lever cannot be moved as far in reverse, which prevents excessively high reverse speeds.

The *mechanical power* from the engine is converted to *hydraulic power* by a pump-motor team as shown in Fig. 38. This hydraulic power is then converted back to mechanical power at the transmission, which is similar to a regular transmission except it doesn't have a reverse gear, because the hydrostatic drive is reversible.

The hydrostatic drive functions as both a clutch and a transmission. The final gear train is usually simplified because the hydrostatic unit supplies infinite speed and torque ranges as well as reverse speeds.

The pump of the unit may be mounted near the engine so that it is driven directly from the engine crankshaft or powershaft. In this design, hoses connect the pump with the motor and the motor is mounted on the transmission housing.

On other combines the pump and motor are mounted on the transmission housing as a unit. The flow of hydraulic oil is direct between the pump and motor—no hoses are required. In this case the pump may be driven by a belt drive from the engine.

The hydraulic system for the drive unit is separate from the regular combine hydraulic system. The schematic in Fig. 39 shows the basic hydrostatic system.

OPERATION OF HYDROSTATIC DRIVE

Here is a simplified description of how a hydrostatic drive operates.

In a hydrostatic drive, several pistons are used to transmit power — one group in the *pump* sending power to another group in the *motor* (Fig. 40).

The pistons are in a cylinder block and revolve around a shaft. The pistons also move in and out of the block parallel to the shaft as we'll see later.

Fig. 37—Combine With Hydrostatic Drive

Fig. 38—Schematic Of A Complete Hydrostatic Drive

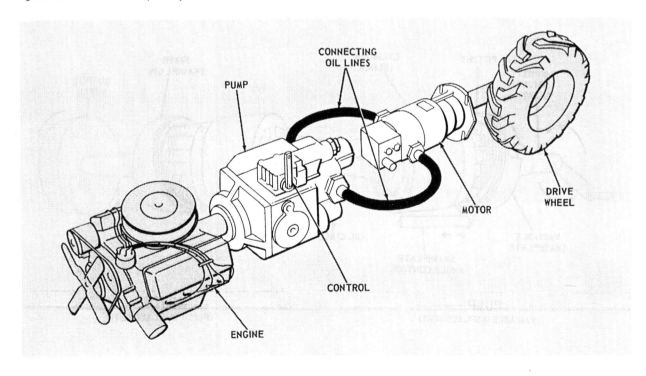

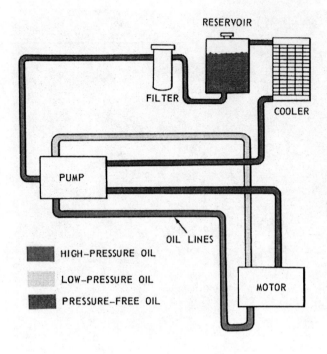

Fig. 39—Complete System For A Hydrostatic Drive

Let's look at one of these cylinder arrangements.

Two cylinders, each containing a piston, are connected by a line (Fig. 41). The cylinders and the line are filled with oil.

When a force is applied to the left piston as shown, that piston moves against the oil. The oil will not compress, so it acts as a solid and forces out the right piston.

One piston for the pump and one for the motor are shown in Fig. 42. To provide a pumping action for the pistons, a plate called a *swashplate* is located in both the pump and motor. The mechanism is designed so the pistons ride against the swashplates.

The angle of the swashplates (Fig. 42) can be varied so that the volume and pressure of oil pumped by the pistons can be changed or the direction of oil flow reversed.

A pump or motor with a movable swashplate is called a *variable-displacement* unit. A pump or motor with a fixed swashplate is called a *fixed displacement* unit.

A variable displacement pump driving a fixed displacement motor is shown in Fig. 40.

As the pump pistons rotate, they move across the sloping face of the swashplate, sliding in and out of their cylinder bores to pump oil in and out. The more the pump swashplate is tilted, the more oil it pumps with each piston stroke and the faster it drives the motor.

The motor swashplate is at a fixed angle so that the stroke of its pistons are always the same. Thus its speed of rotation cannot be changed except as it is driven faster or slower by the pump oil.

Fig. 40—Typical Hydrostatic Pump And Motor Schematic

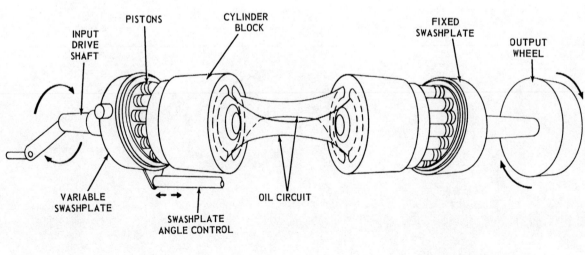

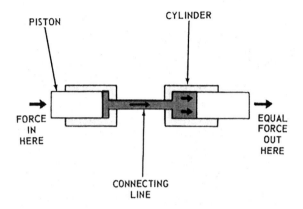

Fig. 41—Two Cylinders Connected

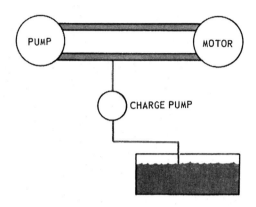

Fig. 43—Pump-Motor Form A Closed Hydraulic Loop

The point to remember now is that a given volume of oil forced out of the pump at a given pressure will cause the motor to turn at a given speed. More oil will increase the flow and speed up the motor; less oil will reduce the flow and slow it down.

The pump is driven by the machine's engine and so is linked to the speed set by the operator. It pumps a variable stream of high-pressure oil to the motor depending on engine speed and tilt of the pump swashplate.

Since the motor is linked to the drive wheels of the machine, it gives the machine its travel speed.

Only three factors control the operation of a hydrostatic drive:

- **Rate** of oil flow—gives the speed
- **Direction** of oil flow—gives the direction
- **Pressure** of oil—gives the power

Control of these three factors is infinite, giving endless selections of speed and torque in a hydrostatic drive.

Fig. 42—Two Connected Cylinders With Swashplates

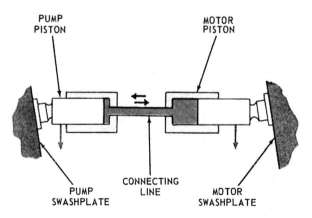

The pump-motor team is the heart of the hydrostatic drive, although the complete hydraulic system (Fig. 39) also includes a reservoir to supply the oil, a filter to remove dirt, and a cooler to remove excess heat from the oil.

Basically, however, the pump and motor are joined in a closed hydraulic loop (see Fig. 43); the return line from the motor is joined directly to the intake of the pump, rather than to the reservoir. The charge pump simply supplies the oil, drawing it from the reservoir.

Fig. 44—How Reversing Is Done

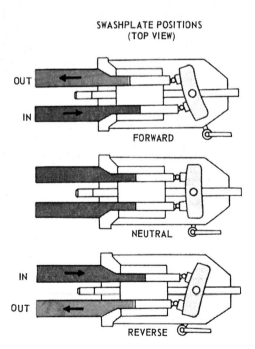

Fig. 45—Four Wheel Drive Provides Extra Traction In Muddy Fields

Fig. 46—Oil Flow Schematic Of Four Wheel Drive System (Forward Shown)

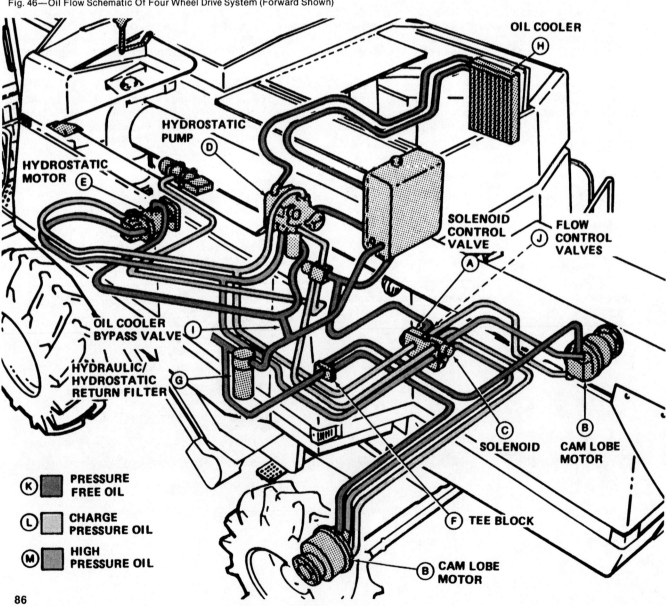

Fig. 47—Four Wheel Drive On Combine

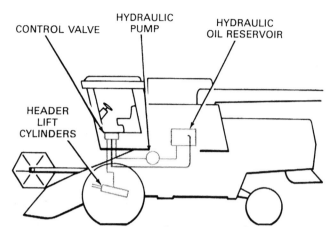

Fig. 48—Early Self-Propelled Combines Had a Simple Hydraulic System

NOTE: *For basic theory of hydrostatic drives, see publications listed in Suggested Readings section in back of this manual.*

The direction of output shaft rotation can be reversed by shifting the pump swashplate over center (Fig. 44).

In NEUTRAL, the swashplate is vertical and no oil is pumped.

In FORWARD, the swashplate is tilted and oil is pumped as shown at the top.

In REVERSE, the swashplate is tilted the opposite way and the unit pumps oil in the opposite direction.

FOUR WHEEL DRIVE SYSTEM

Four wheel drive offers extra traction in muddy fields as shown in Fig. 45. When the combine needs additional traction, the operator turns on a switch which actuates this drive.

The four wheel drive system (Fig. 46) consists of an electro-hydraulic control valve, two fixed-displacement rear wheel motors of a planetary gearing or cam lobe design, an oil filter and connecting hydraulic lines. Hydraulic power for this system is taken from the hydrostatic drive to combine front wheels.

To engage the four wheel drive, the operator sends an electrical signal from the console control switch to the four wheel drive control valve. This valve responds by opening to allow flow of hydrostatic fluid to the rear wheel motors (Fig. 46). The wheel motors convert this hydraulic energy into mechanical energy through a gear reduction to the wheel rims, propelling the combine.

HYDRAULIC SYSTEMS

On early self-propelled combines, the only hydraulic power used was for raising or lowering the header (Fig. 48). Many of today's combines use hydraulics not only to raise and lower the header, but to control the ground speed, raise and lower the reel, control the speed of the reel, drive the belt pickup, vary the speed of the feeder conveyor, control cylinder speed, swing the unloading auger in or out of position, and steady the machine (hillside combines). Also, as we discussed earlier, hydraulics may be used to propel the combine by a hydrostatic drive.

The basic hydraulic system (Fig. 49) consists of a reservoir for storing the oil, a pump for moving the oil, a control valve for directing the flow of oil, and cylinders or motors for doing the work. In this simplified system, the pump is pumping oil to the cylinder control valve and the motor control valve.

The cylinder control valve in Fig. 49 has been shifted to direct the pressure (red) oil to the cylinder to push the load up. The blue oil in the top of the cylinder is allowed to return to the reservoir.

When the cylinder control valve is shifted in the opposite direction, the oil is directed to the top of the cylinder and the load is lowered.

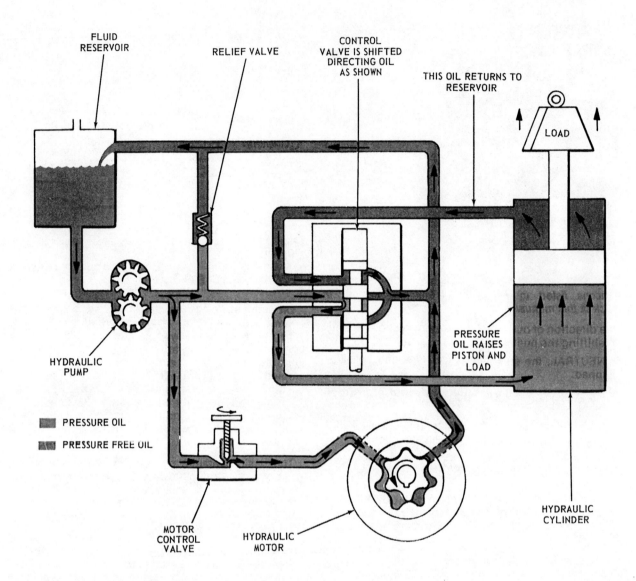

Fig. 49—Basic Hydraulic System

When the control valve is returned to the center position, the oil is trapped in the cylinder at both the top and bottom; thus, the load is held at the position selected.

The motor control valve controls the volume of oil going to the hydraulic motor. As the valve is opened farther, the speed of the motor increases. The oil passing through the motor is also returned to the reservoir. To stop the motor, the operator shuts off the control valve.

When both the control valves have shut off the flow of oil, the oil returns to the reservoir through the center passage of the valve. The relief valve opens when the pressure exceeds safe limits, such as at the end of the hydraulic cylinder stroke.

Several more hydraulic components may be added to this system and sometimes more than one system is used on a combine. A typical complex hydraulic system is shown in Fig. 50 and discussed below.

Reel Lift Cylinders

The operator actuates the control valve to either raise or lower the reel to the desired height. In this design, one cylinder is the master cylinder which controls the slave cylinder so that both move the same distance to keep the reel level.

When the proper height is reached, the control valve is allowed to return to center and the cylinders are held in position by trapped oil.

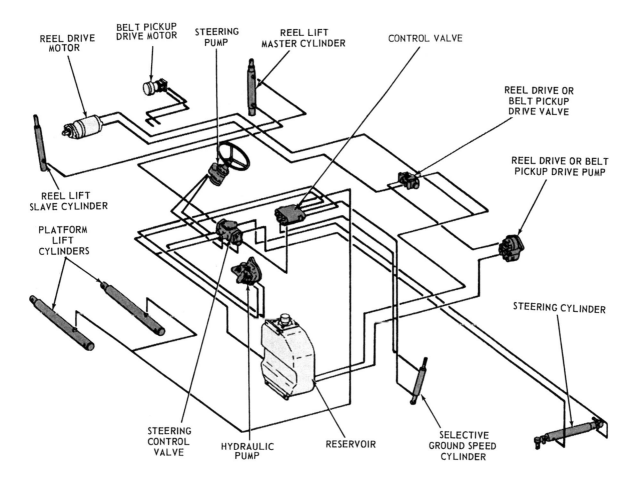

Fig. 50—Typical Hydraulic System On A Modern Combine

Drive Motors

The operator controls the speed of the reel drive or belt pickup drive by opening or closing the motor drive valve. The operator can adjust the reel or belt pickup speed to his ground speed to keep the proper speed ratio. Attachments are available for some combines which automatically adjust reel or pickup speed in proportion to ground speed.

Header Lift Cylinders

Again, by operating the control valve, the operator can raise or lower the header to the desired operating height. When the control valve is returned to center, the header is held in position by trapped oil.

Steering Pump

The steering system in Fig. 50 uses a steering pump which is attached to the steering wheel. When the operator turns the steering wheel in the direction he wants to go, the steering pump delivers oil to the steering control valve which directs the pressurized oil from the main hydraulic pump to the proper end of the steering cylinder. This causes the rear wheels to turn in the desired direction.

Selective Ground Speed Cylinder

The selective ground speed cylinder changes the effective diameter of the variable speed sheaves on the combine propulsion drive. When the operator wants to change ground speed in a particular gear, he operates the control valve to activate the ground speed cylinder to move the variable sheaves to the selected position.

LEVELING SYSTEM (HILLSIDE COMBINE)

Hillside combines are designed to keep the chaff and grain handling parts level while the combine is used on a hillside incline. Specially designed electro-hydraulic leveling systems are used to keep the combine on a horizontal plane. Leveling is activated by two basic methods. They are:

- **Micro Switch Activated**
- **Mercury Switch Activated**

MICRO SWITCH ACTIVATED

The oldest design is the micro switch leveling system (Fig. 51). Micro switches mounted on a leveling control switch box sense the movement of an arm that is attached to the diaphragm piston. For example, when the combine left side wheels become lower than the right, the liquid in the piston shifts (Fig. 52). As the micro switch closes the electrical circuit, the solenoid operated control valve directs oil flow to the top of the left side leveling cylinders. Oil flows out of the right side cylinder and returns to the reservoir.

On gradual slopes, up to 5 or 6 degrees, only a low speed micro switch is actuated. As the slope becomes greater, a high speed micro switch closes and the hydraulic system levels the combine at a faster pace.

When the left-hand cylinder extends and the right-hand cylinder retracts, the fluid gravity booster piston is pushed down in the fluid container. This action, along with the return of the liquid head to its normal height, causes the diaphragm to return to neutral, breaking the electrical circuit and allowing the leveling control valve spool to center. Oil is trapped in both leveling cylinders and the separator is held level while the wheels conform to the contour of the slope.

When leveling in the opposite direction, the change in the liquid head is reversed, which causes the diaphragm in the switch box to move upward. The arm attached to the diaphragm piston closes the upper micro switches, actuating the solenoid and moving the leveling valve spool in the opposite direction. This causes an opposite flow of oil in the leveling cylinders.

Combines can be leveled from front to rear as well as side to side.

Fig. 51—Leveling System For Hillside Combine

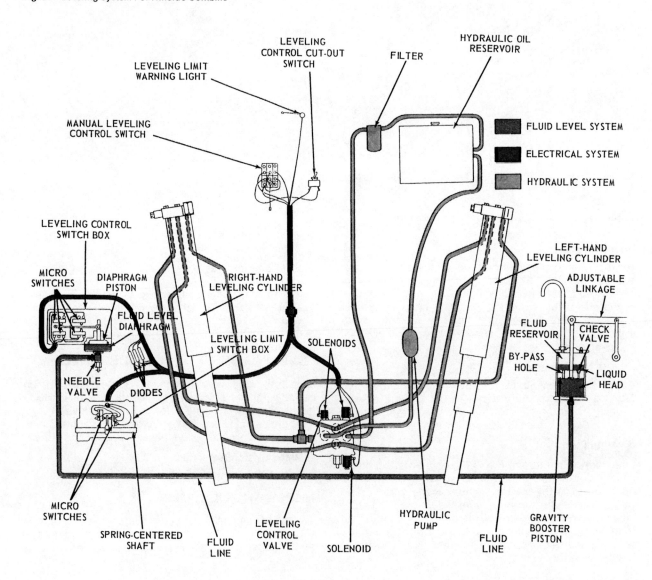

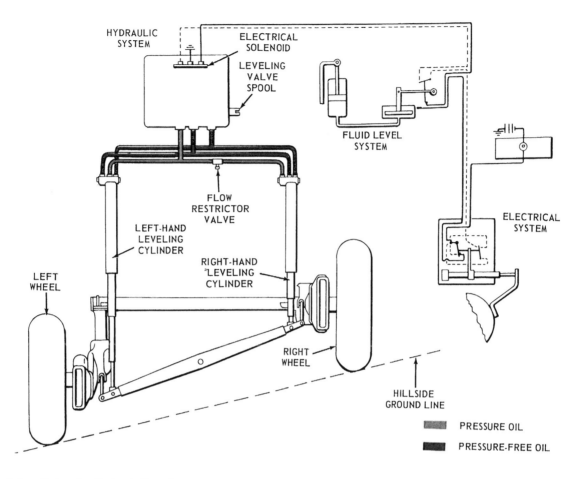

Fig. 52—Automatic Leveling System In Operation

MERCURY SWITCH ACTIVATED

The mercury switch method also is an automatic control. The use of this type of control eliminates the need for the diaphragm piston, and micro switches in the leveling control switch box. These parts are replaced by a mercury switch panel (Fig. 53). As the combine enters the slope of a hill, a stream of mercury starts to run to the low side of the panel.

When the combine is approximately 2 degrees from level, mercury fills the first low angle switch and closes the contact. The solenoid that controls the leveling cylinders is activated. If the angle is more than 5 or 6 degrees, mercury flows to and closes the high angle switch. This switch opens the high hydraulic oil flow solenoid port and the combine levels at a quicker rate.

When the cylinders level the combine, they are held in check until the angle of the combine again changes. The solenoid ports close and trap hydraulic fluid in the cylinders.

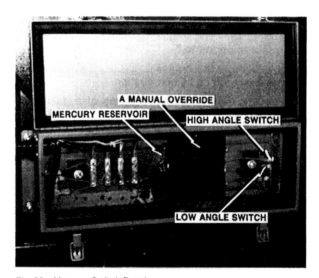

Fig. 53—Mercury Switch Panel
A — High Angle Switch
B — Low Angle Switch
C — Mercury Reservoir

Mercury can be added to or taken out of the mercury reservoir. More mercury will activate the leveling system quicker.

Manual Controls

Both micro and mercury switch automatic controls can usually be overridden by a manual level switch in the operator's cab. This manual control can be used to begin changing the angle of the combine **before** the combine enters an abrupt change in the terrain. Once manual overrides are de-activated, the combine leveling system will return to the automatic mode if it is properly set in the operator's station. The manual override also can be used to tilt the combine for certain maintenance and repair operations.

 CAUTION: Be sure to follow your operator's manual for proper blocking techniques before you work on a machine that is raised on hydraulic cylinders.

CHAPTER QUIZ

1. What are the three power systems required to operate a modern combine?

2. What are the three types of engines used on self-propelled combines?

3. List the following engine strokes in proper sequence:

    A. Compression Stroke   C. Intake Stroke

    B. Exhaust Stroke   D. Power Stroke

4. Name five of the six engine systems.

5. (Fill in blank.) The _____ atomizes and mixes the fuel with air to provide a combustible mixture for the engine.

6. True or false? "The engine lubricating system has only four major parts."

7. (Fill in blank.) The engine _____ system absorbs heat created by combustion.

8. Name the three electrical systems of a gasoline engine.

9. What are the three basic drive systems found on most combines?

10. (Fill in blanks.) The drive systems in question 9 are driven by the _____ countershaft.

11. What four components will the hydraulic system on modern combines control?

12. What are the three systems which operate the leveling system on a hillside combine?

4
Operating Controls

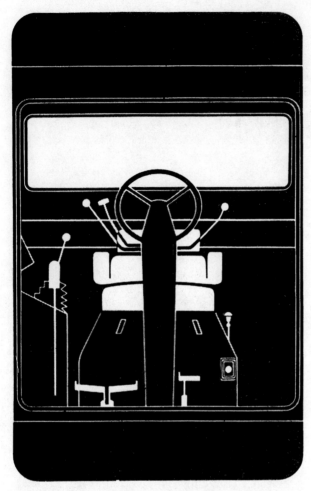

Fig. 1—Most Controls Are Within Easy Reach Of The Operator

Fig. 2—Some Controls Have Different Shapes and Colors To Help Identify Them

INTRODUCTION

Almost all the operating controls and instruments are located at the operator's station within easy reach from the driver's seat (Fig. 1). The operator's station is located high on the front of the combine to give the operator a good view of the header. Here the operator can control all of the combine functions.

Most combines have essentially the same controls and instruments. Their names and locations may be different but they operate the same functions. For example, the control to raise or lower the header may be called either a platform lift control, a table lift control, or a header lift control.

Levers, knobs or switches may be used to control the various components. Some manufacturers have controls of different colors and shapes to help the operator quickly identify the controls while operating the combine (Fig. 2).

Here are some of the color codes one manufacturer uses for the controls:

RED—Combine Movement Controls (throttle, gearshift, selective ground speed control)

YELLOW—Auxiliary Power Controls (separator control, cylinder speed control, and header drive control)

BLACK—Miscellaneous Function Controls (header lift control, reel lift control, etc.)

IDENTIFYING CONTROLS

Each combine operator's manual has a section on controls and instruments to help the operator identify and operate the controls. Because each manufacturer locates these controls in different places, the operator must become familiar with the controls before attempting to operate the combine. Typical controls and instruments on a modern combine are shown in Fig. 3; some combines have only a few of these and others may have different ones. Study this illustration to become aware of how many controls there are and what their functions are.

Fig. 3—Controls and Instruments of a Typical Combine

1—Accumulator Gauge
2—Steering Column Tilt Pedal
3—Starting Aid Button
4—Turn Signal Lever
5—Horn Button
6—Start Switch
7—Brake Pedals
8—Gearshift Lever
9—Hydrostatic Drive Control Lever
10—Header Lift Switch
11—Reel Lift/Feederhouse Speed Switch
12—DIAL-A-MATIC™ Switch
13—DIAL-A-SPEED™ Control Switch
14—Header Engage Switch
15—Separator Engage Switch
16—Manual Reel Speed Switch
17—Reel Fore/Aft Switch
18—Park Brake Switch
19—Cylinder Speed Switch
20—Concave Spacing Switch
21—Lighter
22—Cleaning Fan Speed Switch
23—Four-Wheel Drive Switch
24—Engine Speed Switch
25—Unloading Auger Swing Switch
26—Unloading Auger Engage Switch
27—Seat Height Lever
28—Seat Position Lever
29—Reverser Control Pedal

Before we discuss controlling combine movement and operation, combine and engine *break-in* should be mentioned. Most manufacturers recommend that certain procedures be followed to assure proper break-in of a new combine. These procedures may include operating the combine at reduced loads for a time to seat the piston rings and give the other parts a chance to adjust to loads. Other things, such as checking torque on important bolts, checking stretch of drive belts, and changing oil in the hydraulic system and engine may be recommended to be made after the combine has operated for several hours.

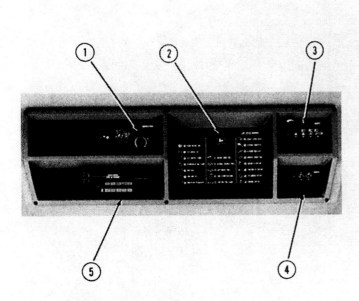
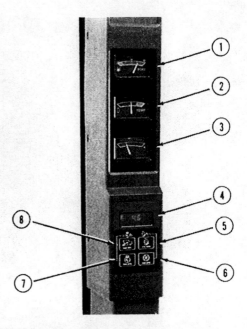

Fig. 3A—Overhead Monitor And Control Panel

1—HARVESTRAK™ Control
2—Warning Display Panel
3—Light Controls
4—Wiper Controls
5—Radio

Fig. 3B—Cornerpost Monitor Panel

1—Fuel Gauge
2—Engine Temperature Gauge
3—HARVESTRAK™ Monitor Gauge
4—Digital Display
5—Cylinder RPM
6—Fan RPM
7—Engine RPM
8—Ground Speed

CONTROLLING COMBINE MOVEMENT

CAUTION: For your own safety, always keep in mind — a combine is a big, heavy machine which requires careful handling at all times. To control combine movement, the operator must know how to:

1. *Operate the engine*
2. *Drive the combine*

The operator must also know how to *transport the combine*, which we will cover later.

OPERATING THE ENGINE

Before attempting to operate the engine, read the operator's manual to become aware of the controls and instruments related to starting and stopping the engine and the safety instructions. Following are some common engine controls and instruments found on modern combines.

Engine Starting Key Switch

The key switch on most combines is similar to those used on automobiles. Three or four positions may be available: "OFF, ON, START, and ACCESSORIES" (Fig. 4). Most manufacturers recommend that you turn the key to the "ON" position before attempting to start the engine to check the operation of the alternator and oil pressure indicator lights. Then turn the switch to "START" to engage the starter to crank the engine. After the engine has started, release the key. The alternator light should go out; if it doesn't go out after 10 seconds, shut off the engine at once and determine the cause.

The operation of other warning lights may also be checked by turning the switch to the "ON" position; they are the parking brake indicator light, oil pressure indicator light, water temperature indicator light, etc.

Engine Choke (Gasoline or LP-Gas Engines Only)

The choke is used to help start the engine when the engine is cold (Fig. 5). Most manufacturers recommend that you pull the choke all the way out when starting. After the engine is started, push the choke

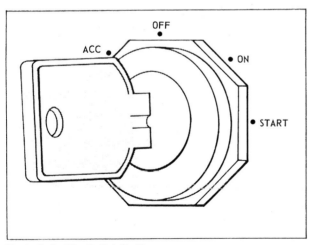

Fig. 4—Typical Key Switch Positions

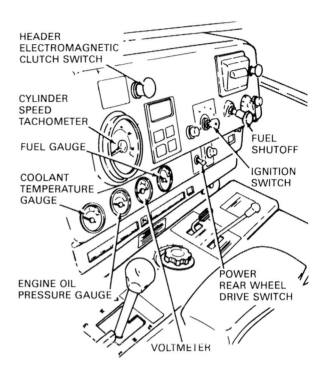

Fig. 5—Typical Engine Controls and Gauges

in slightly until the engine runs the smoothest. As the engine warms, gradually push in the choke until the engine has reached operating temperature. Then push the choke in all the way.

Engine Speed Control

The speed control is usually positioned about one quarter to one third open when starting the engine (Fig. 5). Place it in the full-open position when operating the combine in the field so that the units of the combine are operating at the proper speed, as regulated by the governor.

Fuel Shutoff (Diesel Engines Only)

The fuel shutoff control cuts off the fuel supply to the fuel injection pump (Fig. 5). Usually, the procedure used to stop a diesel engine is to run the engine at slow speed until the engine cools, then close the speed control, shut off the fuel supply, and turn off the key switch.

Alternator Indicator Light

This light glows when the alternator is not charging the battery (Fig. 5). If the light goes on while the engine is running, stop the engine and determine the cause without delay.

Some combines have an ammeter gauge rather than a warning light. This gauge indicates the amount of charge or discharge. If, while the engine is running, the gauge indicates that the system is discharging, stop the engine immediately and determine the cause.

Engine Coolant Temperature Gauge

This gauge indicates the coolant temperature in the engine cooling system — not the quantity (Fig. 5). The gauge usually has a warning zone to indicate when the coolant is above normal operating temperature. If the gauge indicates that the coolant is in the danger zone, stop the engine and determine the cause.

Air Restriction Indicator

The air restriction indicator is usually located on the air intake tube to the engine. The red signal in the indicator appears whenever the air cleaner element is dirty and needs servicing. Most manufacturers recommend that this indicator be checked daily.

Engine Oil Pressure Gauge

This gauge (indicator light on some combines) indicates the pressure of the engine lubricating oil—not the amount of oil in the crankcase (see Fig. 5). When the engine is running, the gauge should show oil pressure in the normal range (the indicator light should not be on). If the oil pressure drops below the normal range and into the danger zone (indicator light on), stop the engine immediately and determine the cause.

Fuel Gauge

The fuel gauge simply tells approximately how much engine fuel is in the fuel tank (Fig. 5).

Engine Tachometer

The tachometer usually shows the engine speed in hundreds of rpm. Many operators use the tachometer to tell if the engine is running at the proper working speed especially when operating in the field; if the engine speed drops appreciably, you must reduce the load by reducing ground speed. Sometimes this gauge contains the hour meter also as shown in Fig. 5.

Hour Meter

The hour meter indicates the accumulated engine service hours (see Fig. 5). This is usually based on the normal operating speed of the engine (2200 to 2900 rpm). The hour meter provides a means of determining when lubrication and periodic services are needed.

Fig. 6—Typical Controls

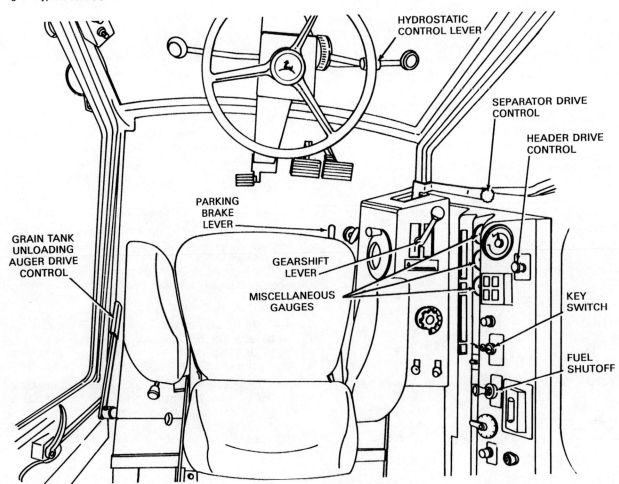

STARTING THE ENGINE

Here we will cover typical starting procedures for gasoline engines, diesel engines, and turbocharged diesel engines. Read the starting tips.

 CAUTION: Follow these procedures to avoid safety hazards and possible damage to the combine.

These instructions are of a general nature. Always follow specific instructions in the operator's manual for the combine being operated.

NOTE: See instructions describing general operation of controls and instruments later in this chapter.

STARTING THE GASOLINE ENGINE

1. Disengage header drive, separator drive and grain tank unloading auger drive (Fig. 6).
2. Place the gearshift lever in neutral.
3. Depress the clutch pedal fully or place the hydrostatic control lever in neutral.
4. Move the speed control one-quarter to one-third open.
5. If the engine is cold, pull the choke control all the way out.
6. Turn the key switch to "ON". Check operation of the warning lights according to the operator's manual.
7. Turn the key switch to "START". After the engine starts, release the key and push the choke in until the engine runs the smoothest.
8. Release the clutch slowly.
9. Be sure the oil pressure gauge registers normal pressure and the warning lights are not on. *If either of these conditions is not correct, stop the engine and determine the cause.*
10. Allow the engine to run for several minutes at fast idle—no load—until the engine and transmission have reached a safe operating temperature.

STOPPING THE GASOLINE ENGINE

1. Depress the clutch pedal or place the hydrostatic control lever in neutral and move the gearshift lever into neutral (Fig. 6).
2. Set the speed control at half speed and allow the engine to run at this speed for several minutes before stopping. This cools the engine and prevents possible damage of parts because of overheating.
3. Close the speed control and turn the key to "OFF".
4. Remove the key from the switch to prevent tampering and unauthorized operation.
5. Engage the parking brake and follow any other parking procedures that may be recommended in the operator's manual.

STARTING THE DIESEL ENGINE

1. Disengage header drive, separator drive and grain tank unloading auger drive (see Fig. 6).
2. Place the gearshift lever in neutral.
3. Depress the clutch pedal fully or place the hydrostatic control lever in neutral.
4. Move the speed control lever to slow idle position and push in the fuel shutoff.
5. Turn the key switch to "ON". Check operation of the warning lights according to the operator's manual.
6. Turn the key switch to "START". After the engine starts, release the key.
7. Release the clutch slowly.
8. Be sure the oil pressure gauge registers normal pressure and the warning lights are not on. *If either of these conditions is not correct, stop the engine and determine the cause.*
9. Allow the engine to run for several minutes at slow idle—no load—until the engine and transmission have reached a safe operating temperature.

STOPPING THE DIESEL ENGINE

1. Depress the clutch pedal or place the hydrostatic control lever in neutral and move the gearshift lever into neutral (see Fig. 6).
2. Set the speed control at half speed and allow the engine to run at this speed for several minutes before stopping. This cools the engine and prevents possible damage to parts because of overheating.
3. Move the speed control to the closed position.
4. Pull out the fuel shutoff and turn the key to "OFF".
5. Remove the key from the switch to prevent tampering and unauthorized operation.
6. Engage the parking brake and follow any other parking procedures that may be recommended in the operator's manual.

STARTING THE TURBOCHARGED DIESEL ENGINE

1. Repeat steps 1 through 8 as described for *Starting the Diesel Engine* (see Fig. 6).
2. Idle the engine for several minutes at low speeds (see operator's manual) *to insure turbocharger lubrication before accelerating or applying load.*

IMPORTANT: If the engine dies when operating under load, immediately restart the engine to prevent overheating of turbocharger parts, which is caused when the flow of oil for cooling and lubrication is stopped.

Here is another important procedure for turbocharged engines:

IMPORTANT: When starting the engine after the com-

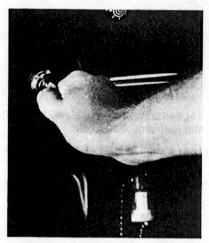

Fig. 7—Before Starting Engine, Make Sure No One Is Standing Near Combine

bine has not been used for an extended period of time or after the oil filter has been changed, pull the fuel shutoff all the way out and crank the engine until the engine oil pressure light goes out. This assures that the turbocharger has a supply of lubricating and cooling oil. Do not operate the starter for more than 30 seconds at a time. After the indicator light goes out, move the throttle to the slow idle position. Make sure the fuel shutoff is all the way in and then start the engine.

STOPPING THE TURBOCHARGED DIESEL ENGINE

1. Depress the clutch pedal or place the hydrostatic control lever in neutral and move the gearshift lever into neutral (see Fig. 6).
2. Allow the engine to idle a few minutes to cool the engine and turbocharger. (Lubrication and cooling of the turbocharger and some engine parts is provided by the engine lubrication oil. Therefore, sudden stopping of a hot engine may allow some parts to overheat and cause possible damage.) Allow the temperature gauge to drop well into the normal range on the dial.
3. Move the speed control lever to the closed position.
4. Pull out the fuel shutoff and turn the key to "OFF".
5. Remove the key from the switch to prevent tampering and unauthorized operation.
6. Engage the parking brake and follow any other parking procedures that may be recommended in the operator's manual.

STARTING TIPS

Before Starting:

Check crankcase oil level.

Check cooling system fluid level.

Check hydraulic system oil level.

Be sure fuel tank is full.

 CAUTION: Before attempting to start the engine, be sure no one is near the combine because a bystander might be injured by moving parts. Be especially concerned for the safety of children.

 CAUTION: Never start or operate the combine in a closed building. Always be sure there is plenty of ventilation when running a combine because the exhaust fumes are poisonous.

WHEN STARTING THE ENGINE, never hold the key in start position for more than 30 seconds at a time or you may damage the starter by overheating internal electrical parts. If the engine does not start within 30 seconds, allow at least 2 minutes for proper cooling of the starter before attempting to start the engine again.

Be sure to pause long enough after a *false start* to make certain the starter has stopped completely before attempting to start the engine again. Otherwise, you may damage the gears on the starter or flywheel.

On DIESEL ENGINES do not let the fuel tank run dry. If it does, you must bleed the entire fuel system to remove air bubbles. Follow instructions in your operator's manual for this procedure.

 CAUTION: Do not attempt to start a combine by towing it. It is an unsafe practice which could result in injury to you or others. Use booster batteries if the engine is hard to start and follow the directions in the operator's manual for proper procedure.

On HYDROSTATIC DRIVE combines, the pump and motor may be damaged if towing is used in an attempt to start the engine.

DRIVING THE COMBINE

To drive the combine safely, the operator must be aware of all the functions of the combine and how they operate; he must know how to operate and adjust the following:

- **Operator's Seat**
- **Steering**
- **Brakes**
- **Propulsion Units**
- **Ladder**

In addition to these, he must know how to *transport* the combine. This will be covered later.

OPERATOR'S SEAT

The operator must be comfortable and within easy reach of the controls of the combine to drive it properly. Before attempting to operate the combine, adjust the seat to your height and reach.

Fig. 8—Driving The Combine

All combines have adjustable seats which can be raised and lowered or moved forward and rearward. Some may be folded up or moved back out of the way so that the operator can stand while driving. A typical seat mounting is shown in Fig. 9.

Fig. 9—Operator And Passenger Seats

To adjust the seat:
1. Sit down and then disengage the seat lock.
2. Put the seat forward or rearward.
3. When the desired position is reached, release the lock and the seat is held in position.

To raise or lower the seat:
1. Remove the spring locking pins.
2. Raise or lower the seat to the desired position.
3. Reinsert the pins to hold the seat in place.

STEERING

Most combines also have an adjustable steering column for individual arm lengths or for standing position. Adjust the steering column by depressing the lock pedal and moving the column to the desired setting (Fig. 10).

The wheels which steer a combine are mounted on the *rear* of the machine. *Be careful when making a turn near obstacles because the rear of the combine may swing around and strike something.*

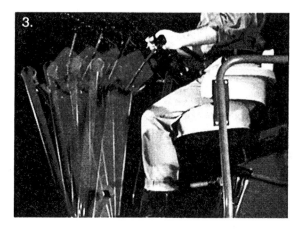

Fig. 10—Adjusting Steering Column

Most steering on modern combines is power steering, which makes driving of the combine down rows or fields an easy task.

BRAKES

The brakes on most combines are the hydraulic type; however, some combines still have mechanical brakes. Usually, two brake pedals are provided which control individual drive wheels (Fig. 11). When used separately, these pedals can be used to assist in turning. When used together, a quick stop is assured.

When stopping the combine, press on both brake pedals. Uneven application of brakes will cause the

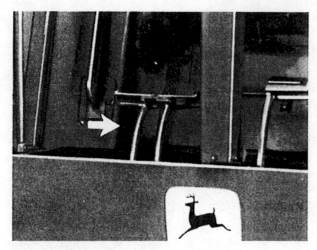

Fig. 11—Brake Pedals

combine to swerve to one side at high speeds. This could result in the combine turning over.

When using the brakes to make sharp turns, slow down to a safe speed. Begin turning the steering wheel *before* applying brake to assist turning. Otherwise the rear wheels will skid sideways and the turn will be more difficult. *Be careful when making a turn near obstacles because the rear of the combine may swing around and strike something.*

 CAUTION: Always reduce travel speed before applying brakes. Quick stops can result in the combine nosing forward.

Some combines have a brake pedal lock which couples the pedals together to aid in even braking when transporting the machine.

Parking Brake

The parking brake locks the wheel brakes so the combine cannot move when left unattended. Never at-

Fig. 12—Operating Parking Brake

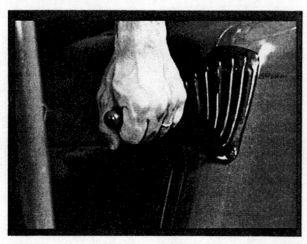

Fig. 13—Propulsion Unit Controls

tempt to move the combine with the parking brake engaged. Some combines have a warning light which tells when the parking brake is engaged.

To engage the parking brake shown in Fig. 12, pull the lever upward; to disengage, push the lever downward.

PROPULSION UNITS

Controlling the propulsion units on *belt drive combines* consists of operating the clutch, gearshift lever, and selective ground speed control.

Controlling the propulsion units on *hydrostatic drive combines* consists of operating the gearshift lever (1, Fig. 13) and the hydrostatic speed range lever (2, Fig. 13) (also see Fig. 14).

Belt Drive Combines

Drive belts and a variable speed sheave assembly transmit engine power to the transmission. Move the selective ground speed control lever and the variable

Fig. 14—Hydrostatic Speed Range Lever

sheave assembly moves, causing the ground travel speed to increase or decrease within a selected transmission gear.

To shift gears:

1. Depress the clutch pedal fully.
2. Then shift into the desired gear.
3. Release the clutch slowly to avoid abrupt starts.

To change ground speed in the gear selected:

1. Move the selective ground speed control lever forward to increase speed.
2. Move the lever rearward to slow down.
3. Release the lever when the desired speed is reached and the hydraulic valve and lever will return to neutral while the travel speed remains as selected.

A ground speed indicator is available for most combines. It has a numbered scale which provides a reference number (not miles per hour) which allows the operator to return to the same ground speed after stopping or changing speed.

Hydrostatic Drive Combines

The hydrostatic unit (pump and motor) transmits engine power to the transmission. When the speed range lever is moved, oil moves through the hydrostatic pump and motor, causing the ground travel speed to increase or decrease within a selected transmission gear.

To shift gears:

1. Place the speed range lever in neutral (Fig. 14).
2. Then shift the gearshift (1, Fig. 13) into the desired position.

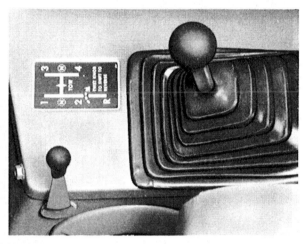

Fig. 16—Gearshift In "Tow" Position

To move the combine forward:

1. Push the speed range lever forward.
2. To increase forward travel in the gear selected, move the lever farther forward.

To move the combine in reverse:

1. Shift the transmission into the desired gear.
2. Pull the speed range lever rearward.

LADDER

Some combines have a movable ladder which can be moved out of the way of uncut grain to avoid grain loss by ladder impact. The ladder may be folded or pivoted out of the way. The ladder shown in Fig. 15 is a pivoting ladder.

Fig. 15—Operating Ladder

Fig. 17—Controlling Field Operation

To move the ladder:

1. Pull lever "A" down to release the lock.

 CAUTION: Keep steps and walking surfaces free of grease and dirt. Use handrails for safe mounting and dismounting. Be careful on frosty mornings. Don't store tools and equipment on operator's platform.

TRANSPORTING THE COMBINE

A combine can be transported by driving it under its own power, carrying it on a truck or towing it. Many combines have a TOW position (Fig. 16) on the gearshift pattern. When towing the combine, place the gearshift in "TOW". Place it in neutral if no tow position is present. Check the operator's manual for recommended towing procedures.

HOW TO TRANSPORT

NOTE: See instructions describing operation of controls elsewhere in this chapter.

1. When transporting the combine under its own power, couple the brake pedals together if a lock is provided.
2. Reduce the width of the combine by folding the unloading auger back along the separator.
3. Remove the header if necessary.
4. If the header and feeder house are removed, wire the hydraulic cylinders up or support them with chains.
5. If the header is left on, raise it to a position which allows good visibility for the operator yet provides ample ground clearance.
6. To reduce the spread of noxious weed seeds, thoroughly clean the combine before leaving one field and going to the next. Sweep trash and straw from the outside of the combine. Open doors at the bottom of the elevators. Remove the grain tank drain hole cover, and run the combine until all straw, trash and grain are removed from inside. Shut off the engine and clean out the interior of the combine.
7. Use appropriate warning devices when traveling on the road or highway, such as a slow moving vehicle emblem, reflectors, operating lights and warning lights. Keep these items clean so that they are visible to other traffic.
8. Have a car or truck drive in front and behind the combine to warn traffic on public roads. Stay in your lane. Pull off, stop, and let traffic pass. Do not drive on the shoulder.
9. Because wheels for steering are in the back, self-propelled machines often fishtail when turned quickly at transport speeds. Steering to the right will whip the rear to the left, and vice versa. The back of the combine will swing out into the path of oncoming traffic.
10. Slowing or braking the combine too rapidly could cause loss of some steering control (weight on rear wheels). This is most noticeable when driving with a corn head, or some other heavy header, or with header raised too high. In this case, most of the weight will be on the drive wheels. Use rear weights. Keep header as low as possible. Use variable speed drive or engine speed control to slow the machine. Reduce speed before applying brakes, and always transport with brake pedals locked together.
11. Check your local governmental regulations for required warning devices.

Fig. 18—Header Drive Electromagnetic Clutch Switch

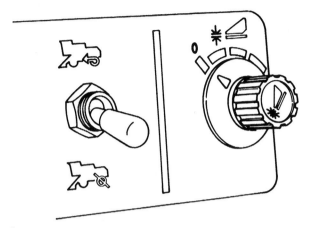

Fig. 19—Header Height Control

105

CONTROLLING COMBINE FIELD OPERATION

Here we will discuss the controls necessary to operate the combine in the field. *Field adjustments* will be covered in the next chapter.

The controls for the following units will be discussed:

- **Header**
- **Separator**
- **Threshing**
- **Cleaning**
- **Grain Tank Unloading**
- **Hillside Combine Leveling**
- **Electrical**
- **Operator's Cab**

HEADER CONTROLS

The combine may be equipped with either a cutting platform, pickup platform, or a corn head. Some of the same controls are used to operate these units.

These controls are necessary to allow the operator to adjust the header for crop and field conditions. It

Fig. 20—Hydraulic Lift Reel Control

Fig. 21—Pickup Platform

106

is important that he control the header for proper cutting, gathering and feeding the crop to the combine separator for efficient harvesting.

Here are the two main header controls.

Header Drive Control

In Chapter 3 we discussed the drive for the header. The header drive may be controlled either mechanically with a lever or electrically with a switch (Fig. 18) and electromagnetic clutch. In either case the operator can engage or disengage the drive easily from his seat.

To engage the header drive with the switch control, simply actuate the switch. To disengage, turn the switch off.

Header Height Control

The operator can control the height of the header, depending on the combine, with either a lever or switch which controls a hydraulic valve that controls the hydraulic cylinders to raise or lower the header to the desired position.

Accumulator

Accumulators are used in the hydraulic system of headers that are both automatic height controlled and those that ride on skid plates. Accumulators help protect the hydraulic system from internal damage by absorbing load and shock as the headers are operated.

An accumulator for a combine is usually a container that is charged with nitrogen gas on one side of a rubber bladder or piston and hydraulic oil on the other side of the bladder or piston (Fig. 22).

The nitrogen gas compresses as pressure from the hydraulic fluid is applied to the bladder or piston. By doing this, the nitrogen absorbs the pressure that is otherwise absorbed by the other parts of the hydraulic system.

Accumulators can be adjusted in two points. The header movement and free play can be stiffened by adding nitrogen gas to the container. (Do not overfill the container.) Another adjustment is at the valve that controls the hydraulic fluid flow into the container. A fully open valve will allow looser header movement than one that is closed.

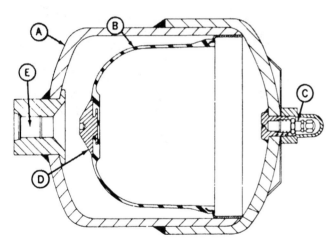

A—Housing
B—Bladder
C—Gas Valve
D—Button
E—Oil Port

Fig. 22—A typical gas filled bladder-type accumulator that is used on a hydraulic operated combine header system

TO RAISE THE HEADER

1. Pull the lever rearward until the header reaches desired height (Fig. 19).
2. Release the lever and the header will remain at the selected height.

TO LOWER THE HEADER

1. Push the lever forward until the header reaches the desired height.
2. Release the lever and the header will remain at the selected height.

Now we will discuss controls for each type of header.

Cutting Platform

Controls for the cutting platform consist of the above controls in addition to hydraulic lift reel and, on some combines, variable speed reel controls. The hydraulic lift reel control lever (Fig. 20) actuates a control valve which causes the hydraulic cylinders for the reel to raise or lower the reel to the desired position. The control is operated just like the header height control above. The variable speed reel control knob allows the operator to adjust the reel speed to the ground speed.

To change reel speed, turn the knob in the proper direction to either increase or decrease speed.

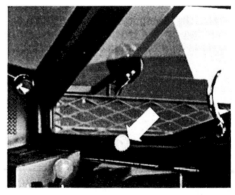

Fig. 23—Separator Drive Control Lever (Overhead View)

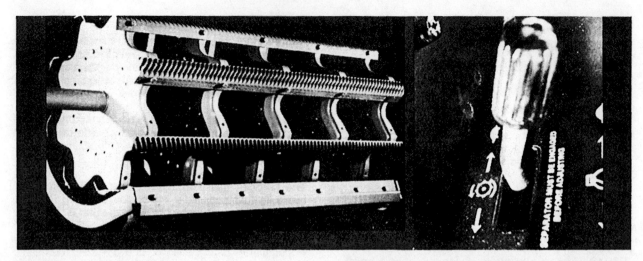

Fig. 24—Cylinder Speed Control

Pickup Platform

Controls for the pickup platform consist of the header drive control and height control as described above. In addition to these, some combines have a hydraulic drive variable speed belt pickup. In some cases the control that is used to control the variable speed reel is used to operate this drive. Again, the operator can choose the exact speed the pickup belt revolves by turning the knob in the proper direction.

Corn Head

The corn head controls also consist of a header drive control and header height control. Many corn heads (Fig. 22) have variable speed drives which are hydraulically controlled. On some combines the same control lever used to raise and lower the reel is used to actuate the hydraulic cylinder which controls the variable drive on a corn head. The operator can adjust the corn head speed to the forward travel of the combine.

Proper operation of snapping rolls, snapping bars, and gathering chains is important for the smooth flow of materials into combine. Improper functioning of these parts can cause uneven feeding, plugging, more down time, and increased risk for the operator.

⚠ **CAUTION: Trying to unclog snapping rolls without shutting off the machine is a leading cause of serious injuries during corn harvesting.**

Fig. 25—Cylinder Speed Tachometer

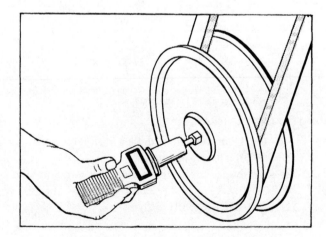

Fig. 26—Checking Cylinder Speed with a Speed-Counter Device

SEPARATOR CONTROL

The separator control lever controls all the drives of the separator. The separator drive must be engaged to operate the following drives:

1. *Header Drives*
2. *Threshing Drives*
3. *Separating Drives*
4. *Cleaning Drives*
5. *Grain Handling Drives (except unloading auger drive)*

Only a few of the above drives have individual controls. Most combines have a separator drive control lever as shown in Fig. 23. To engage the separator drive, push the lever forward. To disengage the separator drive, pull the lever rearward. Other combines may have slightly different operating procedures; refer to the operator's manual.

THRESHING CONTROLS

These are probably the most important controls on the combine. The operator must adjust the cylinder speed and concave spacing to match the crop and conditions in which he is harvesting. Most combines have cranks or ratchets to adjust these components.

Cylinder Speed

On the combine shown in Fig. 24, the cylinder speed is controlled by a ratchet. Cylinder speed may be adjusted from 150 rpm to 1500 rpm. The ratchet controls a chain which changes the settings of the variable speed sheaves driving the cylinder.

To increase the cylinder speed, move the ratchet toward "FAST."

To decrease the cylinder speed, move the ratchet toward "SLOW."

IMPORTANT: Most combine operator's manuals recommend that the cylinder speed be adjusted only when the cylinder is running.

Many combines have a cylinder tachometer which tells how fast the cylinder is running (Fig. 25). Other combines may not have this feature and a speed-counter device must be used to determine cylinder speed (Fig. 26).

Concave Spacing

The combine shown in Fig. 27 has a crank to control concave spacing. The crank controls a gear sector

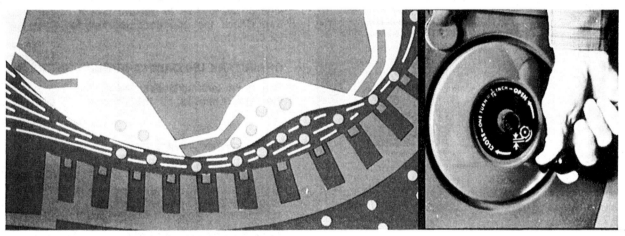

Fig. 27—Concave Spacing Control

Fig. 28—Fan Speed Control

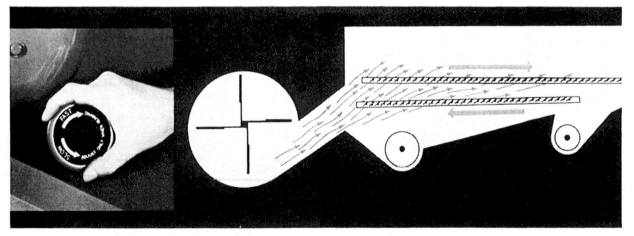

located near the concave. The gear sector is calibrated with reference numbers which tell the operator how wide the concave spacing is. Some concave controls are adjusted so each complete turn of the crank moves the concave 1/16 inch (1.6 mm). With this method the operator must make an initial setting for the crop to be harvested. Then he can make minor adjustments from the operator's station.

To operate this crank, loosen the locking knob first. Then turn the wheel toward "OPEN" or "CLOSED" to adjust the concave spacing. Tighten locking knob.

CLEANING CONTROLS

The cleaning controls adjust the fan speed, chaffer and sieve. The fan must be adjusted to the crop and feed rate. The chaffer is adjusted to allow the fan blast to separate the chaff from the grain and not allow too much coarse material through. The sieve is adjusted to allow only grain through.

To adjust the fan, engage the separator control lever with the combine running. Turn the wheel until the indicator shows the proper fan speed (Fig. 28). This controls the adjustment of the variable speed sheaves which drive the fan. On some combines, fan speed can be adjusted from the operator's cab. On other machines, the operator must leave the cab to change speed. Also, a constant fan speed is used on some combines, with air flow changes made by adjusting fan openings. Refer to the operator's manual for settings.

IMPORTANT: Most fans must be adjusted only when the separator is running.

To adjust the chaffer, push levers "A" to the left to open and push the levers to the right to close it (Fig. 29).

To adjust the sieve, push lever "B" to the right to open the sieve and push the lever to the left to close it (Fig. 29).

When setting the chaffer and sieve, measure the openings of the louvers to get the proper adjustment (Fig. 30). Check the operator's manual for initial settings for each crop.

GRAIN TANK UNLOADING AUGER CONTROLS

The grain tank unloading auger has two operations that are controlled by the operator: 1) swinging the

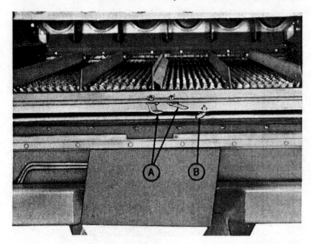

Fig. 29—Chaffer and Sieve Controls

Fig. 30—Measuring Louver Openings

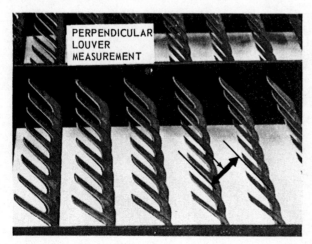

Fig. 31—Grain Tank Unloading Control

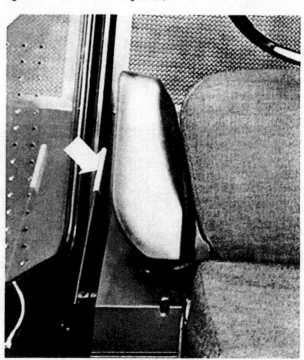

auger into operating position or transport position and 2) emptying the grain tank.

Unloading Grain Tank

To empty the grain tank on the combine shown in Fig. 31, pull the lever rearward to engage the auger drive.

When the grain tank is empty, push the lever forward to disengage the drive.

Positioning the Unloading Auger

Some combines have a hydraulic swing control (see Fig. 33) for the unloading auger and others must be swung into position manually.

To fold the auger back into transport position, release the latch and operate the hydraulic swing control or push the auger into position (Fig. 32). Lock the auger in position with the locking pin.

To place the auger in unloading position, remove the locking pin. Operate the hydraulic swing control or push the auger into position. Latch the auger in position.

An unloading auger extension is required on some combines when using very wide headers to permit easier unloading.

HILLSIDE COMBINE LEVELING

Hillside combines have two controls for the leveling system: automatic leveling and manual leveling.

AUTOMATIC LEVELING is just that; when the operator has the automatic system on, the combine levels itself (Fig. 34). A leveling limit warning light on the control panel warns the operator when the combine has reached the maximum safe leveling limit.

The use of a mercury switch panel also is on automatic control. As the combine enters the slope of a hill, a stream of mercury starts to run to the low side of the switch panel.

Fig. 32—Folding Auger Back

Fig. 33—Placing Auger in Unloading Position

Fig. 34—Automatic Leveling Controls

Fig. 35—Manual Leveling Controls

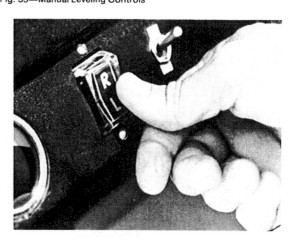

When the combine is approximately 2 degrees from level, mercury fills the first low angle switch and closes the contact. The solenoid that controls the leveling cylinders is activated. If the angle is more than 5 or 6 degrees, mercury flows to and closes the high angle switch. This switch opens the high hydraulic oil flow solenoid port and the combine levels at a quicker rate.

When the cylinders level the combine, they are held in check until the angle of the combine changes again. The solenoid ports close and trap hydraulic fluid in the cylinder.

 CAUTION: When the leveling limit warning light warns that the combine has reached the safe leveling limit, proceed with the utmost caution or the combine may overturn.

MANUAL LEVELING is controlled by the operator with a two-way switch (Fig. 35). If the leveling system fails to function or the operator desires to tilt the separator while on level ground, the leveling mechanism can be controlled with the switch. Push the switch to "R" for right-hand tilt and "L" for left-hand tilt.

IMPORTANT: Do not run the engine for any length of time with the combine in the tilted position on level ground because the oil in the engine crankcase will not be fed properly to the bearings and other moving parts in the engine. The engine could be seriously damaged.

ELECTRICAL ACCESSORIES

The combine may be equipped with the following electrical devices for operating the combine:

- Field and Highway Safety Lighting
- Truck Signal Horns
- Straw Walker Warning Device
- Shoe and Walker Losses Monitoring Unit
- Slow Speed Shaft Monitoring Unit

We will now look at these accessories in detail.

Fig. 36—Field and Highway Lighting

FIELD AND HIGHWAY SAFETY LIGHTING

Most combines are equipped with headlights, field operating lights, taillights and flashing warning lights for highway operation.

Fig. 37—Light Switch

Fig. 38—Truck Signal Horn Button

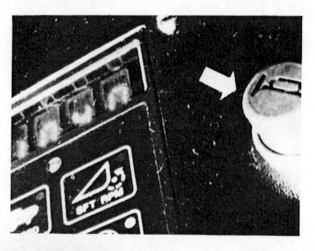

Fig. 39—Straw Walker Plugging Warning Device

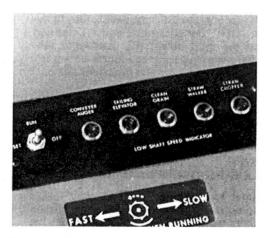

Fig. 40—Slow Shaft Speed Monitor

Fig. 41—The Operator's Cab Provides Comfort In Dust, Heat and Cold

A light switch on the instrument panel operates these lights (Fig. 37). To operate the light switch: turn to the first detent for "OFF"; the second detent (W), flashing warning lights; the third detent (R), headlights, taillight and flashing warning light; the fourth detent (F), headlights, auxiliary headlights and grain tank lights.

TRUCK SIGNAL HORNS

To call the truck to the combine when the grain tank is full, sound the signal horns by pushing the horn button on the instrument panel (Fig. 38).

STRAW WALKER PLUGGING WARNING DEVICE

Some combines are equipped with a straw walker plugging warning device which sounds a horn when the straw walkers become plugged (Fig. 39). When the horn sounds, stop the combine and turn off the engine. Then inspect the straw walkers and remove accumulated straw.

SLOW SHAFT SPEED MONITORING UNIT

This device indicates (with warning lights) when certain "key" shafts of the combine are running below their designed speed (Fig. 40). Units such as tailings elevator, grain conveyor, clean grain elevator, straw walkers and straw choppers can be monitored.

When one or more of the lights glow, the operator must shut off the combine engine and investigate the problem and remedy it before continuing to operate.

ELECTRONIC GRAIN LOSS MONITOR

Electronic measuring and sensing devices are becoming increasingly popular with combine operators. Not only can remote sensing devices help increase yield, they can help save harvest time, and make the job safer.

The most common type of monitoring system is the grain loss monitor. It measures impacts from grain as grain comes off of the cleaning shoe, and straw walker. These

Fig. 42 — The Grain Loss Sensing System In A Typical Combine

A. CONTROL PANEL
B. PREAMPLIFIER CIRCUIT BOX
C. CLEANING SHOE SENSORS
D. STRAW WALKER SENSORS

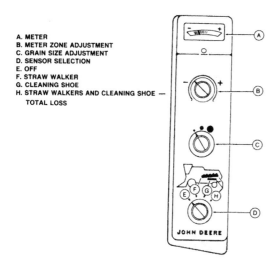

A. METER
B. METER ZONE ADJUSTMENT
C. GRAIN SIZE ADJUSTMENT
D. SENSOR SELECTION
E. OFF
F. STRAW WALKER
G. CLEANING SHOE
H. STRAW WALKERS AND CLEANING SHOE — TOTAL LOSS

grain impacts on the monitor create impulses that are transmitted to a pre-amplifier (Fig. 42). There the impulses are averaged into a DC current that is connected to a read out meter in the operator's cab.

By presetting the grain size, and acceptable loss of grain at the shoe and walker, the combine operator can determine grain loss and modify the combine in order to get the best operating results (Fig. 43).

OPERATOR'S CAB

If the combine is equipped with an operator's cab, it may have a heater, air conditioner, windshield wiper and a radio for the operator's comfort and pleasure.

Here are discussions on the controls of some of these components. Use Fig. 44 for reference.

Fig. 43 — Typical Grain Loss Monitor Control

Fig. 44—Typical Operator's Cab Controls

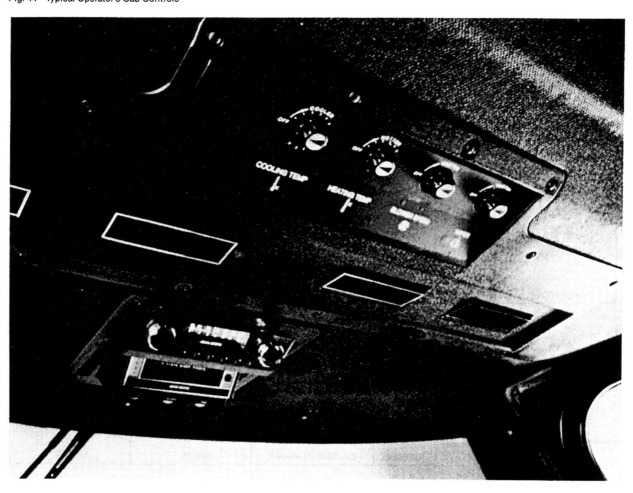

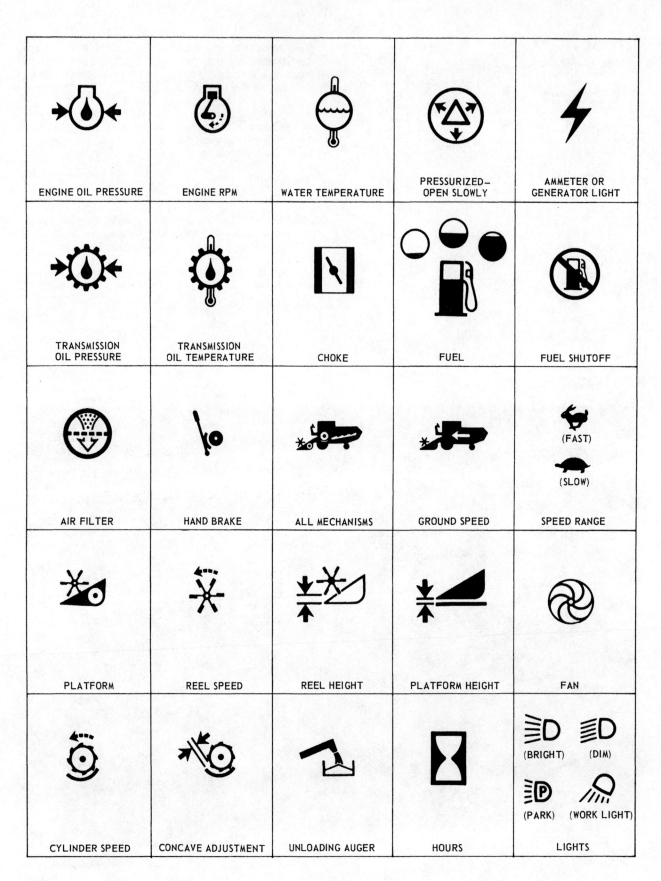

Fig. 45 — Universal Symbols

Heater Temperature Control Switch

To operate the heater, turn the switch knob from the "OFF" position toward the "HOT" position. By turning the switch all the way to the "HOT" position, maximum heating will be provided. The pressurizer fans must be turned on to provide heated air to the operator.

Pressurizer Fans Switch

Most cabs have fans which pressurize the cab. This helps prevent dust from entering the cab. Usually, multiple speeds are available so that the operator can turn the switch to select the most efficient or comfortable air delivery. *The pressurizer fans must be operating whenever the heater or air conditioner is in use.*

Air Conditioner Temperature Control Switch

The air conditioner temperature control switch is a thermostatic-type switch which maintains the desired temperature selected by the operator.

Cool air in the cab is controlled by turning the air conditioner temperature control switch from the "OFF" position toward the "COLD" position. Turning the switch all the way to the "COLD" position will provide maximum cooling. The pressurizer fans must be turned on to provide cooled air to the operator.

High-Pressure Switch Warning Light

This light will glow red when the high-pressure switch has been actuated, indicating it may be necessary to clean the air conditioner condenser.

Condenser Fans Reversing Switch

Occasionally the air flow through the air conditioner condenser can become blocked with chaff and straw.

Some cabs have a reversing switch that will blow air in the opposite direction to blow chaff and straw from the condenser.

This reversing fan should be operated regularly, such as each time the grain tank is unloaded.

Louvered Air Outlets

The air outlets are adjustable so that the operator can control the direction of air flow into the cab.

Other Controls

The cab may be equipped with windshield wipers, a dome light and a radio. Each of the controls for these components is self-explanatory.

SYMBOLS FOR OPERATING CONTROLS

Some manufacturers use symbols instead of words to indicate the function of each operating control. Many of these symbols are shown in Fig. 45. Study these universal symbols and learn to recognize them at a glance.

CHAPTER QUIZ

1. What three indicator lights should glow when the key switch is turned to the ON position?
2. (Fill in blank.) The _____ should be placed in the full open position when operating the combine in the field.
3. True or false? "Gasoline engines should be allowed to cool before shutting them off."
4. Before starting a combine, what four fluid levels should be checked?
5. True or false? "The combine should always be operated at maximum ground speed."
6. What three ways are suggested for transporting a combine?
7. True or false? "The threshing cylinder speed may be changed at any time."
8. What are two of the cleaning unit controls that may need adjustment during operation?
9. True or false? "A warning device may warn the operator if the straw walkers become plugged."
10. True or false? "The operator's cab may contain an indicator which tells when the cab has become pressurized."

5
Field Operation and Adjustments

Fig. 1—Even An Experienced Operator Can Learn New "Tricks"

INTRODUCTION

No one can qualify as a good combine operator by simply reading the operator's manual; a competent operator must have experience. Even those who have had experience can always learn new "tricks." Here we will discuss the basic operation and adjustments which can help the operator do a good job of combine harvesting.

To be a good combine operator, you must know:

1. **Functional Design of the Combine**
2. **Basic Principles of Operation**
3. **How to Make Proper Adjustments**
4. **How to Identify Harvesting Losses**
5. **How to Maintain Efficient Operation**

In previous chapters we have covered the first two items; in this chapter we will discuss the others.

Fig. 2—Poor Harvesting Efficiency in Corn Can Result in a Profit Loss of Nine Percent or More

Fig. 3—Harvesting Efficiency in Soybeans Averages 87 Percent

PROPER OPERATION AND ADJUSTMENT ARE IMPORTANT

Combine harvesting can be profitable only if the operator knows how to adjust the combine properly and operate it efficiently with a minimum of losses.

In corn, for example, recent studies have shown that profit losses of nine percent or more can result if harvesting efficiency is not ideal. In soybeans, the profit loss can be 13 percent or more.

What is ideal efficiency? In some studies the average harvesting efficiency for 100-bushel-per-acre (5 tonnes/ha) corn was 92 percent, while the ideal was 97 percent. In 30-bushel-per-acre (1.6 tonnes/ha) soybeans, 87 percent efficiency was average, while ideal was 97 percent.

How do you get ideal efficiency? This chapter helps solve that complex problem, which is basically a question of *proper operation and adjustment* of the combine.

Harvest losses directly result in decreased profits. These losses are caused by poor adjustment and operation of the combine. In corn yielding 100 bushels per acre (5

Fig. 4—Signs of Poor Combine Harvesting

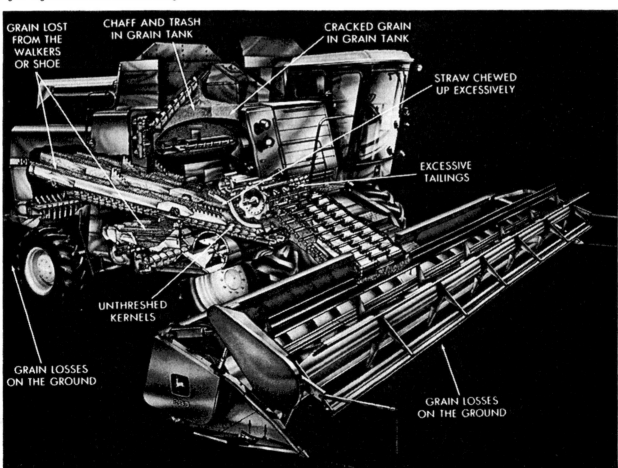

tonnes/ha), for example, if losses were four bushels per acre (200 kg/ha), value lost per acre would be ten dollars at a price of $2.50 per bushel (10 cents per kg). In addition, marketing losses include penalties for low test weight, moisture, damages, and excessive trash or foreign material, all of which can add significantly to the profit losses.

WHAT ARE PROPER OPERATION AND ADJUSTMENT?

The operator must recognize the effects of both good and poor operation. Illustrated in Fig. 4 are some of the **signs of poor combine harvesting:**

1. Grain losses on the ground
2. Unthreshed kernels on the straw, cob or in the pod.
3. Straw chewed up excessively
4. Grain lost from the straw walkers or shoe
5. Excessive tailings in the tailings elevator
6. Cracked grain in the grain tank
7. Chaff or trash in the grain tank
8. Marketing penalties for low-quality grain due to harvesting damage or crop condition

FACTORS WHICH AFFECT COMBINE HARVESTING

Several important factors affect the harvesting of crops with a combine. The following factors must be considered in every instance of combining:

1. Skill of the operator
2. Condition of the crop and field
3. Adjustment of the combine
4. Proper operating speed of harvesting components
5. Ground speed of the combine
6. Width of the header

Fig. 5—Four Harvesting Methods

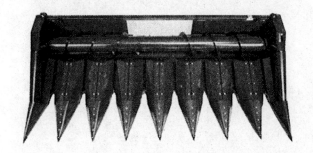

COVERAGE IN THIS CHAPTER

This chapter is not meant as a guide to combine harvesting, but it is meant to explain the principles of combine harvesting which are invaluable to any combine operator.

Here are the subjects which will be covered in this chapter:

- **Planning and Preparation**
- **Preliminary Settings**
- **Field Operation and Adjustments**
- **Field Problems**

PLANNING AND PREPARATION

Prior to combining, the operator must make plans and prepare for the harvesting operation. It is not enough to know how to operate a combine; the operator must also consider the following factors:

1. Harvesting Methods
2. When to Combine
3. Consequences of Early and Late Harvesting
4. Production Capacity of the Combine
5. Mechanical Condition of the Combine

Let's look at these subjects to learn their importance.

HARVESTING METHODS

Usually the crop determines the harvesting method (Fig. 5). For example, when harvesting corn with a

Fig. 7—Typical Pull-Type Windrower

combine, a corn head must be used; when harvesting other crops, a cutting platform is used. A row-crop head may be used for soybeans, sunflowers or sorghum. In addition, some crops are harvested with a windrow pickup platform. Many combine owners raise one or two different crops which may require different headers (discussed in Chapter 2) and often different combines, such as a corn-grain-soybean combine, an edible-bean combine, or a rice combine. These combines may be converted to other crops by changing the header or threshing cylinder or both. Of course, other minor changes may be made.

Fig. 6—Self-Propelled Grain Windrower

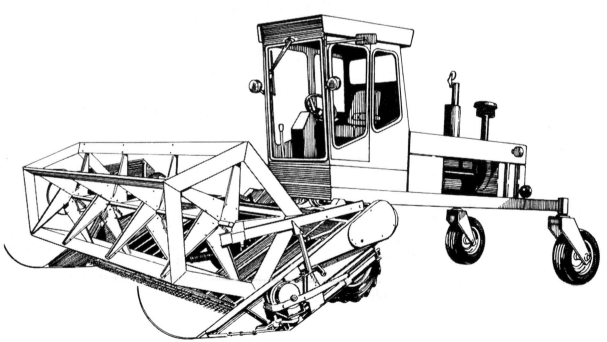

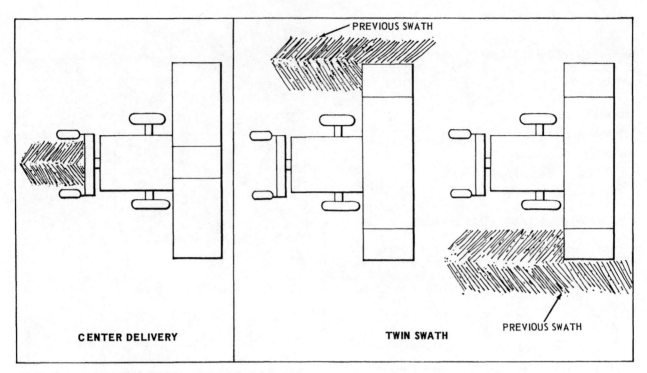

Fig. 8—Two Types of Deliveries Are Possible

One harvesting method which needs further explanation is the *windrow method.* We will discuss the combining of the windrow later in this chapter. However, before a combine can be used to harvest the crop, the crop must be cut and windrowed by a grain windrower.

GRAIN WINDROWERS

Grain windrowers (sometimes called swathers) are often used when there are many weeds in the grain, when there is considerable moisture at harvest time, when the crop ripens unevenly or, when it does not

Fig. 9—Typical Windrower Construction

Fig. 10—The Reel Divides and Delivers The Crop

Fig. 12—Twin Windrower Swath

ripen in time for harvest before bad weather begins (Fig. 6). In these cases it is usually desirable to cut the grain with a windrower and thresh it later with a combine equipped with a windrow pickup attachment (pickup platform or belt pickup).

The grain windrowers place the grain in a windrow. Advantages provided by a windrow are:

- Crop is less susceptible to bad weather
- Allows even drying of grain
- Allows full use of the combine by placing up to 42 feet (12.8 m) of grain in narrow swath.

TYPES AND SIZES

The grain windrower may be either a *self-propelled* (Fig. 6) or a *pull-type* (Fig. 7). The self-propelled machines are gradually replacing the pull-types because they are more maneuverable and they free the tractor to perform other functions.

Fig. 11—Windrower Canvases Discharge The Crop To The Ground

With both the self-propelled and pull-types, two kinds of deliveries are possible. The most common is the center or single delivery (Fig. 8, left). The other type is the twin or double swath (right). When the twin swath is used, windrows are placed side by side. The first swath is shown in blue. With the double swath, one windrow is placed on top of the other as shown.

The center delivery sizes available for grain are 12 to 30 feet (3.7 to 9.1 m). The most common are 16, 18, or 21-foot (4.9, 5.5 or 6.4 m). The twin swath types are available from 18 to 25 feet (5.5 to 7.6 m).

CONSTRUCTION

The grain windrower is constructed with a bat-type reel at the front (Fig. 9). The reel makes the first contact with the grain. A cutterbar operates beneath the reel to cut the crop. Behind the cutterbar is a pair of canvases. They travel from the outside edge of the platform to an opening in the center of the single swath platform.

This opening is usually about three to four feet wide.

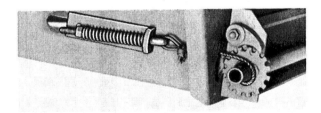

Fig. 13—Canvas Adjusting Mechanism

The opening on a self-propelled machine is in the center. The opening of a pull-type machine is generally offset to the left, to provide clearance for the tractor on the next round. The twin swath platforms have openings on each end (Fig. 8).

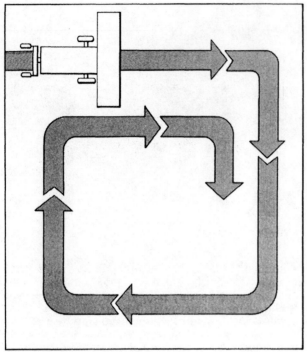

Fig. 14—Opening A Field With Self-Propelled Center Delivery Windrower

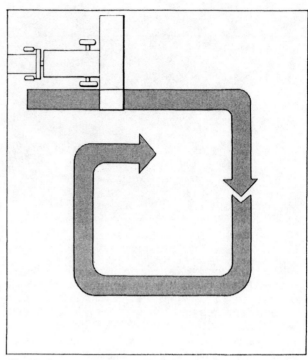

Fig. 16—Opening A Field With A Self-Propelled Twin Swath Windrower

Fig. 15—Opening A Field With A Pull-Type Center Delivery Windrower

Fig. 17—Second Round With a Self-Propelled Twin Swath Windrower

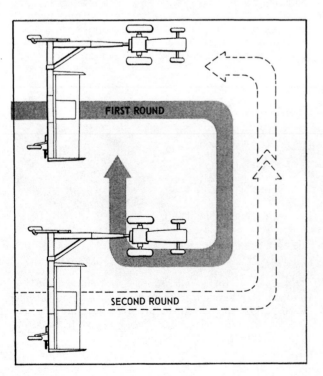

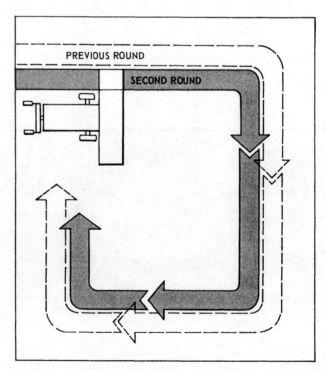

OPERATION OF WINDROWER

The bat reel of the windrower divides and delivers the crop. The material is then cut off by the cutterbar. The reel then delivers the cut crop to the canvases behind the cutterbar. From here the crop is then conveyed by the canvases to the opening and discharged to the ground (see Fig. 11).

The twin-swath platform on self-propelled windrowers has a movable canvas. This canvas is moved by hydraulics to allow delivery to the opening at either end of the platform. With the pull-type windrower, there is an attachment which will cover the opening with a movable canvas which conveys the material to the left-hand side of the machine, beside the previously laid swath (see Fig. 12). This twin swath attachment is also moved by tractor hydraulics to open and close the platform opening for proper swath position.

HOW TO WINDROW GRAIN

The grain is ready to be windrowed when it reaches 35 percent moisture level. This is normally referred to as the late dough stage. This is when the line between the green and gold on the straw stem is about eight inches above the ground. After the grain reaches this maturity, the crop will not be damaged by cutting. This guideline suggests windrowing about one week earlier than many consider grain is ripe enough to cut. University tests show, however, that windrowing does not damage the crop at this stage.

Fig. 18—Opening A Field With A Pull-Type Twin Swath Windrower

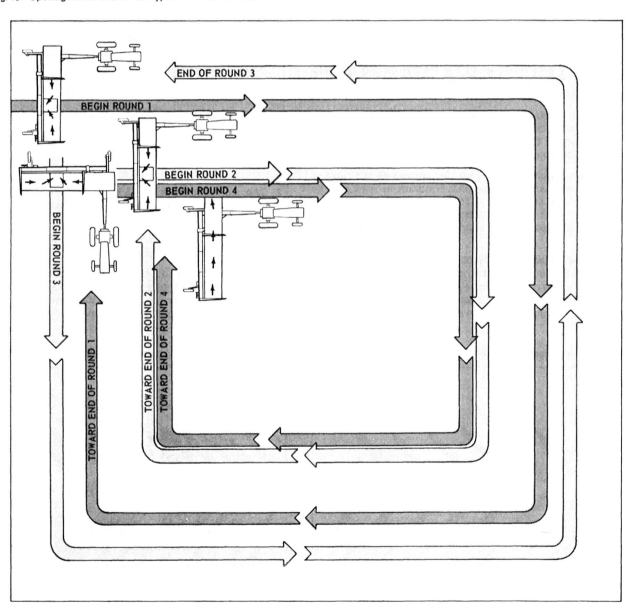

Checking the Machine

Prepare the windrower by thoroughly checking the machine for loose bolts, lubricating properly (refer to the operator's manual) and checking the condition of the cutting parts.

Adjust the windrower before going to the field. Check the condition of the canvas and adjust the tension if necessary. The canvas is usually tightened by a spring-loaded mechanism. Be sure the tighteners are capable of moving freely (see Fig. 13). The drive belts and chains must be inspected for wear and then tightened properly if that adjustment is needed.

Opening The Field

To open a field with a self-propelled center delivery platform, go around the field in a clockwise direction and continue until the field is completely cut (see Fig. 14).

To open a field with a pull-type single swath platform, start in a clockwise direction for a couple of rounds (see Fig. 15). Next, reverse direction to pick up the portion knocked down by the tractor on the opening swath. Then go back to cutting the rest of the field in a clockwise direction.

To open a field for a self-propelled twin swath platform, shift the canvas to the left. This delivers the grain to the opening on the right (see Fig. 16). Go in a clockwise direction for one round.

On the second round, shift the canvas to the right to allow left-hand delivery and continue in a clockwise direction (see Fig. 17). This places the two swaths beside each other. Continue in clockwise direction, alternately placing the windrow to the right and to the left.

To open a field for a pull-type twin swath platform, go in a clockwise direction for two rounds with the canvas in the single swath position (see rounds 1 and 2, Fig. 18). Reverse direction and cut the material knocked down by the tractor on the opening swath (see round 3).

Shift the canvas and start to twin swath beside the second swath laid (see round 4). Then continue to alternately lay the windrow beside and then behind the tractor. (This method may be altered because of tillage practices or because a self-propelled machine may be used to open fields.)

FIELD ADJUSTMENTS

The field adjustments necessary are those required to lay a proper type of swath. The most desirable type of swath is a windrow with most of the grain heads in line, with a slight herringbone pattern.

The drawbacks to the herringbone windrow are:

1. With all the heads in the center of the windrow (Fig. 19) during wet weather, the heavier heads will go down and touch the ground.

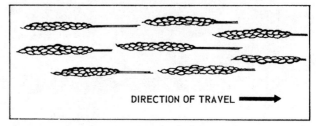

Fig. 20—Heads-In-Line Windrow (Poor Windrow)

2. The combine cylinder will only be used at the center since that is the only position of the heads.

The drawback to a strict heads-in-line windrow is that some straw will go down between the grain drill rows and never be picked up (Fig. 20).

To obtain the desired heads-in-line with a slight herringbone pattern requires proper speed relationships between the reel, the ground and the canvas (Fig. 21). It also requires proper height from the ground to the

Fig. 19—Herringbone Swath With Heads In Center (Poor Windrow)

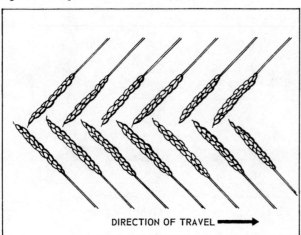

Fig. 21—Heads-In-Line With Herringbone Pattern (Best Windrow)

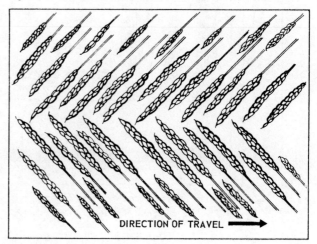

Fig. 22—Stubble Should Be 4 to 8 Inches (10 to 20 cm) High

Fig. 23—Transporting Pull-Type Windrower

cutterhead and from the cutterbar to the reel. These speed and height relationships must be maintained to give the desired windrow.

Start the adjustments by obtaining proper cutterbar height. The stubble should be 4 to 8 inches high (10 to 20 cm) (Fig. 22). This height is enough to allow adequate air circulation and yet not tall enough to allow bending so the heads may touch the ground. The height of the reel from the cutterbar to the reel bat is determined by the height of the grain. Position the reel to let it touch only the top six inches (15 cm) of grain. Continual adjustment of the reel height may be necessary due to uneven height of the standing grain.

Once the height relationships have been set, the reel speed and ground speed must be set. Determine the proper ground speed. This will normally be four to six miles per hour (6.4 to 9.7 km/h) depending on the ground conditions. Next, set the reel speed. The reel speed should be 25 percent faster than ground speed. Finally, set the canvas speed, if adjustable, to lay the desired windrow at the selected ground speed and reel speed settings.

Other aids to forming good windrows are a short piece of canvas at the openings for the butts to fall on and the use of windrow forming rods.

TRANSPORTING THE WINDROWER

The self-propelled windrower should be equipped with flashing warning lights and a SMV emblem (check local regulations). Some pull-type windrowers may be placed in a transport position by removing the wheels to the end to allow endwise transport (Fig. 23). With either type, safety must be stressed due to the width of the machine.

MAINTENANCE

The components requiring maintenance for the grain windrower are the belts, chains, canvases, and cutting parts. Check these items periodically while using the machine. Follow the preventive maintenance procedures as outlined in John Deere's FMO Preventive Maintenance manual.

Storage

Storage recommendations for a grain windrower are:

1. If possible, shelter the windrower in a dry place, or cover it with a tarpaulin.
2. Clean the windrower thoroughly. Chaff and dirt will draw moisture and rust the steel.
3. Remove and clean the canvases. Hang the canvases in a dry place where they will not be subject to damage from inclement weather or rodents.
4. Remove the tension from all belts and clean them thoroughly.
5. Clean the chains thoroughly. Brush heavy oil on the chains to prevent corrosion.
6. Lubricate the windrower completely. Grease the threads on adjusting bolts and the sliding surfaces of the variable-sheave assemblies.
7. Paint all parts from which paint has worn.
8. Support the platform with blocks to level it.
9. Block up the windrower to take the load off the tires. Do not deflate the tires. If the windrower is stored outside, remove the wheels and tires and store them in a cool, dry place away from direct sunlight.

Stock of Parts

It is advisable to have on hand a number of parts which are subject to wear and breakage. The parts list for grain windrowers includes:

Knife Guards	*Knife Holddowns*	*Rivets*
Knife Wear Plates	*Chain Couplers*	*Drive Belts*
Knife Sections	*Engine Air Filters*	*Engine Oil Filters*

FIELD PROBLEMS WITH WINDROWERS

The following chart provides a general guide to solving field problems. Each operator's manual has such a chart to help the operator make proper adjustments or to correct improper operation of windrower.

Problem	Cause	Remedy
CROP LOSS AT CUTTERBAR		
Shattering of grain ahead of cutterbar.	Excessive agitation of grain heads due to incorrect entry of reel slats into crop.	Set reel so reel slats feed material smoothly to cutterbar and canvases.
	Reel speed not coordinated with ground speed, causing excessive agitation before crop is cut.	Change reel drive to coordinate reel speed with ground speed so reel will move material smoothly and evenly.
	Ground speed too fast for condition of crop.	Reduce ground speed so reel will not bat crop, causing shattering of grain heads.
Cut crop building up and falling from front of cutterbar or loss of grain heads at cutterbar.	Reel not adjusted low enough for proper delivery of cut crop to canvases.	Set reel low enough to sweep material from cutterbar to canvases.
	Cutting platform too high, cutting stalks too short for proper delivery.	Lower cutting platform so stalks of crop will be long enough for smooth, even feeding and to support windrow.
POOR CUTTING ACTION		
Ragged and uneven cutting action.	Various parts of cutterbar, such as knife sections, guards, wearing plates, etc., are worn, damaged, or broken.	Check and replace all worn and broken parts on cutterbar to obtain an even cutting of crop.
	Bent knife, causing binding of cutting parts.	Straighten the bent knife. Check guard alignment and align if necessary for a smooth cut.

Problem	Cause	Remedy
	Cutterbar out of register.	Adjust register. See Operator's Manual.
	Knife clips not adjusted to permit knife to work freely.	Adjust knife clips so knife will work freely, but still keep knife sections from lifting off guards.
	Looseness between knife back and guard.	Adjust knife clips so knife back is snug to guard.
	Ground speed too fast.	Slow down.
	Lip of guard out of adjustment or bent causing poor shearing action.	Adjust lip of guard so it is parallel to shear edge of guard.
	Loose cutterbar drive belt.	Adjust drive belt.
Excessive vibration of cutting parts.	Excessive looseness of cutting parts and knife drive.	Remove all excessive play from cutterbar and knife drive to eliminate vibration. After removing excessive play, make certain cutterbar and knife drive are properly adjusted and move freely.
IMPROPER REEL DELIVERY		
Reel wrapping in tangled and weedy crops.	Incorrect location of reel and improper setting of reel slats.	Place reel well ahead and down.
	Reel speed too fast.	Reduce speed of reel to allow weedy crops to fall onto platform.
Reel carrying crop around.	Tall grain or nodding varieties of crops catch on reel slats and arms.	Increase width of reel slats with wire screen or canvas for nodding varieties of crops.
	Reel speed too fast.	Reduce speed of reel. Reel should turn just enough faster than ground travel so that crop

(Continued)

Problem	Cause	Remedy
		heads are laid back on cutting platform.
	Reel height too low.	Raise reel height to reduce amount of crop gathered by reel.
Crop falling in front of cutterbar after it is cut.	Reel speed too slow	Adjust reel to turn 25 percent faster than forward travel of combine.

SUMMARY: WINDROWERS

The grain windrower makes a windrow to be picked up by the grain combine (Fig. 24). It comes in two styles, self-propelled and pull type. Either type can lay two types of windrows — single swath and twin or double swaths.

Fig. 24—The Grain Windrower Makes a Windrow To Be Picked Up By The Combine

Fig. 25—Combine When The Crop Is Ready

WHEN TO COMBINE

Some combine operators may argue about the best time to combine—one man may say that he likes to combine before the crop reaches full maturity because he can get all his crop harvested early—another may like to harvest at the optimum maturity because he will have less crackage and losses.

In any case, if the crop is harvested too early or too late, the crop may suffer damage or losses which could cut profits considerably.

The usual rule of thumb is to *harvest as early as possible when the moisture content has dropped to acceptable levels.*

PREMATURE HARVESTING results in yield losses and a reduction in quality. Immature grain produces

Fig. 27—Corn With 20 to 30 Percent Moisture May Be Harvested

Fig. 26—Delayed Harvesting Results In Lodged and Down Plants

smaller yields and less weight. The grain is also harder to thresh, which tends to cause threshing damage and incomplete threshing.

DELAYED HARVESTING results in losses from shattered grain, lodged plants, down plants, and reduced weight. Shatter losses occur because the grain is too easily threshed. Losses from lodged and down plants occur because the header cannot gather the grain properly. Reduced weight means weight loss which may result in penalties of 1/2 to 1 cent per pound under minimum weight.

BEST HARVESTING TIMES

Best harvesting times vary with the crop and weather. The proper stage for harvesting is when the grain will give the biggest yield at the highest quality.

Here are general suggested harvesting times for corn, soybeans and wheat:

Corn—Usually corn gives its highest yields when the moisture content is between 20 and 30 percent. At higher moisture (35 percent), losses result from cracking and small pieces of kernels (fines). Also, profit is lost from the cost of drying the corn to a moisture content which is safe for storage.

Soybeans—Usually soybeans are harvested when the moisture content is between 12 and 14 percent. When the plants are very dry, gathering losses may be high. This is because the stems and pods become brittle and shatter when they come in contact with the cutterbar and reel. If the bean crop is too dry, it may help to harvest after a light rain or dew. The extra moisture will make the pods and stems more durable and will cut down on shatter losses.

Wheat—Direct combining of wheat is usually done when the moisture content is below 14 percent. At this moisture content, however, shattering and cutter bar losses are greater than at higher moisture levels.

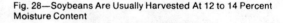

Fig. 28—Soybeans Are Usually Harvested At 12 to 14 Percent Moisture Content

Fig. 29— Wheat Is Usually Harvested When Moisture Content Is Below 14 Percent

At 14 to 20 percent moisture, heads will be harder to thresh from the straw. However, wheat can be harvested at any of the above moisture levels because grain can be dried to safe storage levels. It is a matter of figuring the advantages and disadvantages. At the hard dough stage, wheat can be cut, windrowed and allowed to dry to safe levels.

Other Common Crops—Best moisture levels for harvesting oats, barley, rye and sorghums is usually 13 to 14 percent. If harvested at higher moisture, drying must be used to bring the grain down to a moisture content safe for storage.

PRODUCTION CAPACITIES OF COMBINES

The production capacity of a combine is important to the operator. Sometimes, smaller capacity means less

Fig. 30—Production Capacity Is Important

grain harvested per day and so smaller profits. By estimating the number of acres per hour a combine can harvest, the operator will know how long it will take to completely combine his fields. This will give him a better idea of when he should start, and whether he needs more equipment or help.

The chart in Fig. 31 (next page) shows effective harvesting capacity of combines operating at *harvesting efficiency of 75 percent*, which is considered average. This means that 25 percent of the time is lost due to the following reasons:

1. *Refueling*
2. *Lubricating*
3. *Adjusting machine*
4. *Unloading*
5. *Turning at ends of fields*
6. *Unclogging machine*
7. *Breakdowns*

The percentage of time lost can be more or less than 25 percent according to layout of the field, adjustment of the machine, preventive maintenance practices, and operator's skill.

HOW TO USE THE CAPACITY CHART

To use the chart in Fig. 31: First determine the width of the header. Our example, shown in blue, is a combine with a 16-foot (4.9 m) header. If the combine is equipped with a corn head, figure the width of the header by the row spacing and number of rows the header can harvest (for example, a corn head which is set to combine three 30-inch [76 cm] rows is equivalent to 3 × 30 inches [7.6 × 76 cm] or a header width of 7.5 feet [2.3 m]).

Next, draw a vertical line from the appropriate header width to the line representing miles per hour ground speed (example is 3 mph [5 km/h]). Now draw a horizontal line over to the line representing the time lost. From this point draw another vertical line to the top of the chart to determine the approximate acres harvested per hour (example is 4.25 acres per hour).

Or, assume the header width is 16 feet (5 m) and the combine has a field speed of 4 mph (6 km/h). Draw a vertical line from the header width to the line representing ground speed of 4 mph (6 km/h). Next, draw a horizontal line to the line representing time lost. From this point draw another vertical line to the top of the chart and read the approximate hectares harvested per hour (2.25 ha/h [6 acres/hour] in our example).

Now you know what the combine is capable of in one hour of time. Divide this figure into the total acres to be combined and you will find the number of hours it will take to harvest the crop.

Keep in mind that you may not be able to travel at the speed you selected because the capacity of the combine is affected by the crop and field conditions.

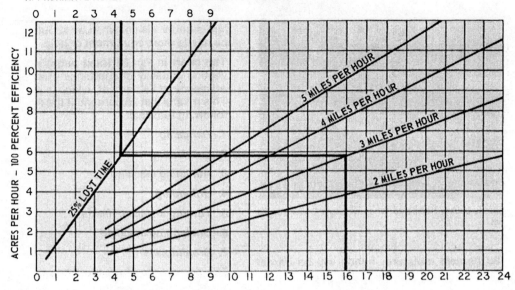

Fig. 31—Combine Harvesting Acres-Per-Hour Chart (See Hectares-Per-Hour Chart On Page 207 Of Appendix)

The average speed most combines travel is between 2.5 and 4.5 mph (4 and 7.25 km/h). This can also vary with the field conditions and the skill of the operator.

HOW TO FIGURE GROUND SPEED

Most combine operator's manuals give approximate ground speed ranges in each gear, but this does not give you accurate miles per hour because of machine load and tire slippage.

Here is a simple procedure for figuring the ground speed:

1. Measure off a distance of 100 feet (30 m) parallel to the combine and in the direction of travel. Mark the distance with stakes.
2. Operate the combine at the ground speed appropriate for the crop and condition.
3. Keep a record of the time it takes the combine to travel the distance between the stakes.
4. Use the simple formula which follows to determine ground speed.

$$\text{Miles per hour} = \frac{\text{distance traveled (ft)} \div \text{travel time in seconds}}{1.466}$$

5. For example, a combine traveled the 100 feet in 23 seconds. Substitute these values in the formula:

$$\text{MPH} = \frac{100 \text{ ft.} \div 23 \text{ secs.}}{1.466} \text{ which equals: } \frac{4.3}{1.466} \text{ (or } 4.3 \div 1.466\text{)}$$

which equals: 2.9 miles per hour

If a combine with a 5 m head traveled 30 m in 18 seconds, speed would be:

$$\text{km/h} = \frac{3.6 \times 30}{18} = \frac{108}{18} = 6 \text{ km/h}$$

6. Consult the chart in Appendix to determine the approximate area each combine can harvest per hour at the ground speed measured. A 16-foot header traveling 2.9 mph would cover 4.25 acres per hour; while a 5 m header traveling 6 km/h would harvest 2.25 hectares per hour — both working at 75 percent field efficiency.

MECHANICAL CONDITION OF COMBINE

One of the costliest factors of combine harvesting is a machine which is in poor mechanical condition. A combine which does not receive good preventive maintenance will cost more in repairs, in addition to time lost due to breakdowns in the field. Poor mechanical condition may also cause poor harvest-performance. Harvesting components which are damaged or worn cannot be adjusted to perform satisfactorily.

A combine should be checked and repaired between seasons to avoid unnecessary downtime when it can be most costly.

Here are some of the problems which should be corrected before attempting to begin the harvesting season:

Fig. 32—Check The Condition Of The Combine Regularly

HEADER COMPONENT PROBLEMS

Cutting Platform Problems (Fig. 33)

1. Bent knife guards
2. Badly worn, broken, or missing knives
3. Worn or broken cutter bar parts, such as hold-down clips or knife guards.
4. Broken reel slats
5. Broken, bent, or missing reel fingers

Fig. 33—Check The Cutting Platform

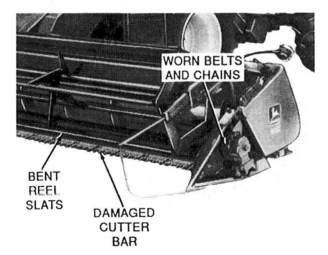

6. Worn or loose chains and belts
7. Worn or bent feeder conveyor chain
8. Bent or missing auger fingers
9. Worn bearings
10. Loose or missing bolts

Corn Head Problems (Fig. 34)

1. Faulty gatherer points
2. Worn or loose gatherer chains
3. Worn snapping plates
4. Worn or broken stalk plates
5. Worn or bent trash knives
6. Worn or bent feeder conveyor chain

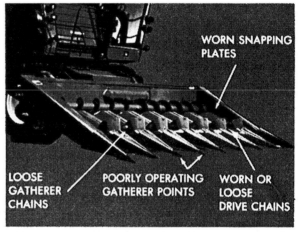

Fig. 34—Check The Corn Head

7. Worn or loose drive chains
8. Worn bearings
9. Loose or missing bolts

SEPARATOR COMPONENT PROBLEMS

Threshing Section Problems (Fig. 35)

1. Worn or bent cylinder or rotor bars
2. Worn or bent concave bars
3. Dirt packed in cylinder, concave, or stone trap
4. Misalignment of cylinder concave
5. Worn or loose drive belts or chains

Separating Section Problems (Fig. 36)

1. Damaged straw walkers
2. Torn or missing straw walker curtain
3. Worn or loose drive belts

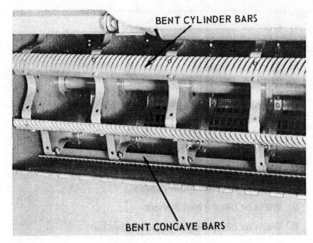

Fig. 35—Check The Threshing Section

Cleaning Section Problems (Fig. 37)

1. Damaged fan blades, or fan shutters (if used)
2. Damaged chaffer sieve
3. Loose or bent shoe hangers

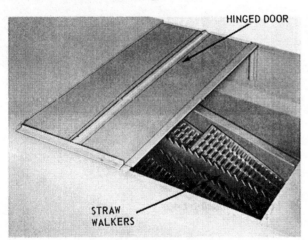

Fig. 36—Check The Straw Walkers

Grain Handling Problems (Fig. 38)

1. Worn elevator chains with missing paddles
2. Badly worn, bent, or broken augers
3. Dirt packed in grain tank or elevators

Other Problem Areas (Fig. 39)

1. Worn or loose drive chains and belts
2. Damaged controls
3. Worn bearings
4. Loose or missing bolts
5. Worn or broken sheaves and sprockets

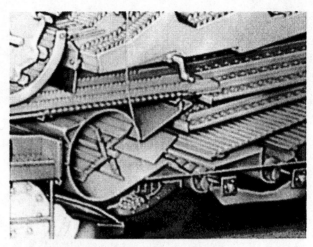

Fig. 37—Check The Cleaning Section

Engine Problems

1. Engine not properly tuned
2. Engine speed too fast or too slow
3. Leaks in air, fuel, or oil lines
4. Dirty air and oil filters
5. Leaking or dirty cooling system

Fig. 38—Check The Elevators And Augers

Fig. 39—Check All Drive Chains And Belts

PRELIMINARY SETTINGS OF COMBINE

All combine operator's manuals have charts which recommend preliminary settings of the machine (Fig. 40). These settings are starting points only; each crop and field condition will require further adjustments. Very few combines will do a good job when adjusted at these starting points unless field conditions are perfect.

Here we will discuss the following important items:

- **Proper operating speed**
- **Suggested combine settings**

PROPER OPERATING SPEED

Every combine separator is designed by the manufacturer to operate at a particular speed regardless of the crop harvested. Never alter the recommended speed or the combine will not perform satisfactorily.

If the speed of the separator is slower than normal, the entire combine will run slower. This will cause sluggishness which will result in clogging and grain losses.

If the speed of the separator is faster than normal, material will pass through the combine too rapidly which will result in grain losses and excessive strain and wear on all components.

Always check the separator speed according to the operator's manual. Most combines are set to operate at the proper speed when the engine runs at maximum speed. If the separator is not running at the proper speed with the speed control in this position, adjustment is necessary.

Each combine has one particular shaft as the starting point for checking operating speed. Some machines use the cylinder beater cross shaft while others use the primary countershaft speed as the

Fig. 40—Read The Operator's Manual For Recommended Settings

starting point. Always refer to the operator's manual to determine how to check the operating speed.

SUGGESTED COMBINE SETTINGS

Each combine operator's manual has lists of suggested machine settings for starting points in each crop to be harvested. These suggested settings are just that—recommended starting points which must be adjusted carefully to the crop and field conditions. Usually a range is given which tells the approxi-

Fig. 41—Checking The Separator Speed With A Speed-Counter Device

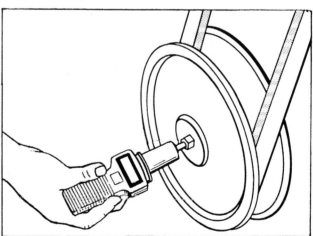

mate settings for average conditions. When setting a particular combine, consult the operator's manual to determine where in the range to begin.

After combining the crop for a while at these settings, check for grain losses and adjust the combine as necessary; *field adjustments* will be covered later in this chapter.

On the following pages we will list typical settings for the most popular crops harvested by combines. Each crop is listed in alphabetical order by its popular name along with an illustration of the harvested crop. While many of these crops may be harvested in other ways for different uses, we will confine this discussion to combine harvesting. Keep in mind that some of these crops are harvested for seed or grain while many varieties are also harvested for hay, silage, and other purposes which require different methods of harvesting. For example, oats may be harvested for seed or grain with a combine, but when it is used for hay or haylage a combine isn't used.

Let's look at some of these crops on the following pages. Notice that typical cylinder, concave, chaffer and sieve settings for a conventional combine are given along with a brief description of the harvesting characteristics. Because cleaning fan settings vary considerably between makes of combines, they are not given here. These settings may vary a great deal with each manufacturer because of the various designs of combines, so always consult the operator's manual for specific recommendations.

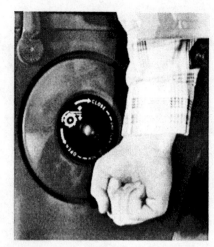

Fig. 42—Adjust Combine To Suggested Starting Points

NOTE: The concave settings shown indicate the range for the front of the concave.

Concave spacing on rotary combines is generally set somewhat wider than on conventional combines but the range may overlap for some crops. And, although the recommended range of rotor speeds for a particular crop may overlap the speeds suggested here, the best rotor speed for any crop will depend on rotor diameter, combine design and crop conditions. So, always consult the combine operator's manual for specific recommendations.

SUGGESTED COMBINE SETTING CHARTS

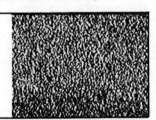

ALFALFA
Cyl. rpm	500-1300
Concave	1/8"-3/8" (3.2-9.5 mm)
Chaffer	3/8"-1/2" (9.5-12.7 mm)
Sieve	1/16"-1/8" (1.6-3.2 mm)

Alfalfa may be combined directly for seeds if the crop is in proper condition. However, when weeds or uneven ripening is a problem, it is usually cut with a windrower and cured before combining.

BARLEY
Cyl. rpm	650-1300
Concave	1/8"-5/8" (3.2-15.9 mm)
Chaffer	1/2"-3/4" (12.7-19 mm)
Sieve	1/4"-1/2" (6.4-12.7 mm)

Barley tends to shatter from the stalk when ripe and under direct combining methods. Shatter losses may be reduced by windrowing the crop when the heads have turned golden yellow and the straw is slightly green. After drying three or four days, it can be combined with a windrow pickup unit.

BEANS
Cyl. rpm	250-700
Concave	1/2"-1" (12.7-24.4 mm)
Chaffer	1/2"-3/4" (12.7-19 mm)
Sieve	3/8"-1/2" (9.5-12.7 mm)

Beans (edible beans) are usually harvested when most of the pods have turned yellow, but before they shatter due to excessive dryness. They may be combined directly or windrowed first. Usually a low cylinder speed and a wide concave clearance are recommended to prevent cracking or splitting the beans.

BUCKWHEAT
Cyl. rpm 650-1300
Concave 1/8"-1/2"
(3.2-12.7 mm)
Chaffer 1/2"-3/4"
(12.7-19 mm)
Sieve 1/4"-3/8"
(6.4-9.5 mm)

Buckwheat usually ripens after the first frost. It is threshed easily by direct combining or it can be windrowed to aid curing and later combined with a windrow pickup unit.

CLOVER
Cyl. rpm 800-1400
Concave 1/8"-3/8"
(3.2-9.5 mm)
Chaffer 1/4"-5/8"
(6.4-15.9 mm)
Sieve 1/16"-1/8"
(1.6-3.2 mm)

Clover seed is usually harvested with the windrow method because of weed and moisture problems. After the windrow is dry and cured, it is combined with a windrow pickup unit. If the seed is combined directly from a standing crop (when 70 to 80 percent of the heads are ripe), the combine header must cut just under the heads because the stems are still green.

CORN
Cyl. rpm 400-900
Concave 1"-1 1/2"
(25.4-38.1 mm)
Chaffer 7/16"-5/8"
(11.1-15.9 mm)
Sieve 1/2"-5/8"
(12.7-15.9 mm)

Corn (field shelled) is harvested with a combine equipped with a corn head. Cylinder filler plates are used to hold the ears close to the concave for complete shelling. These plates also keep cobs from passing through the cylinder unthreshed. Early harvesting is popular, but because the moisture content may be 20 to 30 percent, artificial drying is used to bring it to a safe storage level of 13 to 14 percent.

CORN-COB MIX (Cracked Kernel)
Cyl. rpm 800-1300
Concave 3/8"-1/2"
(9.5-12.7 mm)
Chaffer 3/4"-1"
(19-25.4 mm)
Sieve Not Used

Corn-Cob Mix (Cracked Kernel) is used for feeding purposes. The threshing cylinder is operated at a fast speed which cracks the kernels. The sieve must be removed to allow the pieces of cob to fall through and be carried to the grain tank.

CORN-COB MIX (Whole Kernel)
Cyl. rpm 500-900
Concave 3/8"-1/2"
(9.5-12.7 mm)
Chaffer 3/4"-1"
(19-25.4 mm)
Sieve Not Used

Corn-Cob Mix (Whole Kernel) is also used for feeding purposes. The threshing cylinder is operated at slower speeds to prevent cracking the kernels. The sieve must be removed in this case also.

FLAX
Cyl. rpm 800-1300
Concave 1/16"-1/2"
(1.6-12.7 mm)
Chaffer 5/16"-1/2"
(7.9-12.7 mm)
Sieve 1/16"-3/16"
(1.6-4.8 mm)

Flax is often windrowed before combining, but it can be harvested as a standing crop when thoroughly dry and free of weeds. The best harvesting conditions, however, are produced with the windrower method. Harvesting is usually started after the majority of the bolls are ripe.

CANOLA
Cyl. rpm 450-1000
Concave 3/16"-1/2"
(4.8-12.7 mm)
Chaffer 1/4"-3/8"
(6.4-9.5 mm)
Sieve 1/8"-3/16"
(3.2-4.8 mm)

Canola (Rape) seeds are usually combined directly after the leaves drop off the stalk. The seeds are easily threshed from the pods even when the stems are green. Care must be used to avoid taking too much of the stalk into the combine because the high moisture content will make separation of the seed difficult.

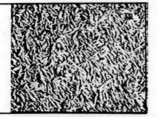

GRASS	
Cyl. rpm	800-1300
Concave	1/8"-3/8"
	(3.2-9.5 mm)
Chaffer	1/2"-5/8"
	(12.7-15.9 mm)
Sieve	1/8"-1/4"
	(3.2-6.4 mm)

Grass Seed (most varieties) may be combined directly, but usually it is better to combine from a windrowed crop. Each method depends on the height of the grass, moisture content and ripeness of the head. When ripe, the seed is easily shattered from the stems. Because of this, the crop is usually cut when the stems are still green.

LESPEDEZA	
Cyl. rpm	650-1200
Concave	1/8"-5/8"
	(3.2-15.9 mm)
Chaffer	1/4"-5/8"
	(6.4-15.9 mm)
Sieve	1/8"-3/8"
	(3.2-9.5 mm)

Lespedeza is often harvested as a standing crop, but it can be windrowed. It should be cut when slightly green to prevent shattering. Otherwise, the straw may break easily and add more material to the cleaning shoe. The fan speed must be low to prevent blowing the light seed out of the combine. This can result in increased tailings which may be unavoidable.

MILLET	
Cyl. rpm	800-1000
Concave	1/16"-3/16"
	(1.6-4.8 mm)
Chaffer	1/2"-5/8"
	(12.7-5.9 mm)
Sieve	1/8"-1/4"
	(3.2-6.4 mm)

Millet seeds are difficult to combine directly because the seeds shatter from the stem before all the seeds are ripe. Therefore, the crop is windrowed before combining with a windrow pickup unit.

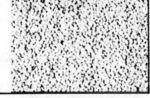

MUSTARD	
Cyl. rpm	750-1300
Concave	1/16"-3/8"
	(1.6-9.5 mm)
Chaffer	5/8"-3/4"
	(15.9-19 mm)
Sieve	1/4"-3/8"
	(6.4-9.5 mm)

Mustard seeds may be combined directly if the pods have matured properly while standing. If not, then the crop is windrowed to cure and combined with a windrow pickup unit.

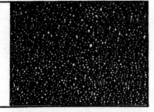

OATS	
Cyl. rpm	700-1300
Concave	1/4"-5/8"
	(6.4-15.9 mm)
Chaffer	5/8"-3/4"
	(15.9-19 mm)
Sieve	1/4"-1/2"
	(6.4-12.7 mm)

Oats are usually combined directly; however, when weeds or uneven ripening is a problem, windrowing is recommended. Windrows allow the green weeds to dry and reduce the moisture of the oats. Then the crop can be combined with a windrow pickup unit.

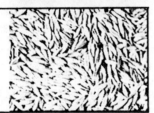

PEAS	
Cyl. rpm	300-550
Concave	1/4"-5/8"
	(6.4-15.9 mm)
Chaffer	5/8"-3/4"
	(15.9-19 mm)
Sieve	3/8"-1/2"
	(9.5-12.7 mm)

Peas (edible) may be either windrowed and combined after curing or combined directly when ripe and dry. If peas are too dry, they will crack easily.

RICE	
Cyl. rpm	700-1050
Concave	1/16"-
	(1.6-12.7 mm)
Chaffer	5/8"-3/4"
	(15.9-19 mm)
Sieve	1/4"-3/8"
	(6.4-9.5 mm)

Rice is harvested by direct combining. It is a difficult grain to thresh because it is hard to strip from the straw. A spike-tooth threshing cylinder is usually used because of its aggressive threshing action. Rice should be threshed as soon as it ripens to avoid crackage by the sun if allowed to stand in the field too long. Rice may often be down or lodged following storms which makes harvesting more difficult.

RYE
Cyl. rpm 800-1325
Concave 1/16"-3/16"
         (1.6-4.8 mm)
Chaffer 5/8"-3/4"
         (15.9-19 mm)
Sieve   1/4"-1/2"
         (6.4-12.7 mm)

Rye can be combined directly or harvested by the windrow-pickup method, depending on conditions. It tends to shatter easier than wheat and may ripen unevenly.

SORGHUM
Cyl. rpm 750-1300
Concave 1/8"-1/2"
         (3.2-12.7 mm)
Chaffer 3/8"-5/8"
         (9.5-5.9 mm)
Sieve   1/4"-1/2"
         (6.4-12.7 mm)

Sorghum (grain) or milo-maize may be harvested directly by a combine. When combining sorghum, the cutting platform must be carried as high as possible to avoid taking in too much of the stalk. The stalk is pithy and damp; if too many stalks are threshed, they will carry seeds out of the combine. The dampness will also add excessive moisture to the seeds.

SOYBEANS
Cyl. rpm 400-850
Concave 3/8"-1"
         (9.5-25.4 mm)
Chaffer 1/2"-3/4"
         (12.7-19 mm)
Sieve   3/8"-1/2"
         (9.5-12.7 mm)

Soybeans are combined directly when the moisture content drops to 12 to 14 percent. Soybeans tend to shatter and crack when ripe; therefore, header and threshing operations must be carefully adjusted to match conditions. If beans are extremely dry it may be advisable to harvest at night when pods become tough from dew to reduce shatter losses.

SUNFLOWER
Cyl. rpm 375-600
Concave 1/2"-1 1/2"
         (12.7-38.1 mm)
Chaffer 1/2"-3/4"
         (12.7-19 mm)
Sieve   1/2"-5/8"
         (12.7-15.9 mm)

Sunflower seeds may be combined directly at 8 or 9 percent moisture content. When combining directly, the cutting platform must be carried as high as possible to avoid taking in too much stalk which will make separating more difficult. Special attachments may be used on the platform.

TREFOIL
Cyl. rpm 950-1200
Concave 1/16"-3/16"
         (1.6-4.8 mm)
Chaffer 1/4"-3/8"
         (6.4-9.5 mm)
Sieve   1/8"-1/4"
         (3.2-6.4 mm)

Trefoil (birdsfoot) seeds may be combined directly or the windrow-pickup method may be used. When combining directly, a desiccant (drying agent) is applied to the crop prior to threshing. The seed shatters easily after the pods are ripe.

VETCH
Cyl. rpm 400-1000
Concave 1/4"-3/4"
         (6.4-19 mm)
Chaffer 1/4"-1/2"
         (6.4-12.7 mm)
Sieve   1/4"-5/16"
         (6.4-7.9 mm)

Vetch seeds are harvested by direct combining or by the windrow-pickup method, depending on the locality and variety. To avoid shattering, it is usually windrowed when the lower pods are ripe.

WHEAT
Cyl. rpm 750-1325
Concave 1/8"-1/2"
         (3.2-12.7 mm)
Chaffer 5/8"-3/4"
         (15.9-19 mm)
Sieve   1/8"-1/4"
         (3.2-6.4 mm)

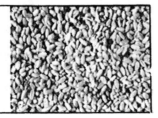

Wheat is primarily combined directly as a standing crop. However, if the crop has excessive weeds and does not ripen evenly, or may be damaged by bad weather before it ripens completely, the windrow-pickup method is used. It is usually harvested when moisture content is below 14 percent.

Fig. 43—Operating In Conditions Such As These Requires A Good Operator

FIELD OPERATION AND ADJUSTMENTS

Operating the combine in the field and making the necessary adjustments requires the skills of a good operator—not just a combine driver. A good combine operator can save several dollars per acre over a combine "driver." In this section we will discuss what a good combine operator must know to do quality harvesting.

OPERATING PROPERLY IN THE FIELD

One important thing that the operator must keep in mind when operating in the field is *the relation between combining speed and crop losses.* Even when the combine is properly adjusted, losses can occur because of excessive speed. The operator must judge what acceptable losses are. Some operators will tolerate more losses and crackage when they want to get the crop harvested faster.

The combine operator's manual is an excellent guide to proper operation and adjustment. Refer to it frequently when problems are encountered because it has the answers to many troubles.

GENERAL RULES FOR GOOD COMBINING

Here are some general practices for good combining:

1. When first beginning to combine a field, get the feel of the combine's ability to handle the crop by operating at a slow ground speed. Use a lower gear than normal, but don't reduce engine speed; if you do, the combine will not perform efficiently. Gradually increase ground speed and check the results until problems are encountered, such as unacceptable losses or threshing damage. Under normal conditions you should be able to operate at ground speeds between 2-1/2 and 3-1/2 miles per hour (4 and 5.6 km/h).

2. Don't hesitate to make adjustments to the combine if they are necessary. Know why you must make the adjustment before doing it. <u>Make only one adjustment at a time and check the results before making any others.</u>

3. Check frequently for proper threshing action and adjust the cylinder speed and concave spacing as necessary.

4. Check for grain losses when checking threshing action. Make adjustments as necessary, such as reducing ground speed, changing cylinder-concave, fan, chaffer and sieve settings.

5. When using a cutting platform in a standing crop, cut as high as possible without missing too many low grain heads. In down and tangled crops, use lifting guards and keep the header low and the ground speed slow.

6. Keep the reel height and speed adjusted for changes in crop height and ground speed.

7. When using a corn head, keep the header low to get low ears. Keep the corn head centered in the rows to prevent stalks from bending and losing ears.

8. Adjust the cleaning units when losses over the shoe or excessive tailings occur.

9. Don't overload the combine by operating at faster ground speeds; losses will increase tremendously.

Fig. 44—When Operating In Weedy Conditions, Slow Down

10. When operating in poor conditions, such as a weedy crop or a hard-to-thresh crop, slow down and make frequent checks for combine performance (Fig. 44).

Now let's look at some of these operations and adjustments more closely. Here we will discuss the following:

- **Operating and Adjusting Cutting and Feeding Units**
- **Operating and Adjusting Threshing Units**
- **Operating and Adjusting Cleaning Units**

Later in this section we will tell how to calculate crop losses and how to reduce them.

OPERATING AND ADJUSTING CUTTING AND FEEDING UNITS

The crop must be cut, gathered and fed properly to the threshing unit or shattering and uneven feeding will result. This in turn will affect the overall efficiency of the combine.

Here we will discuss operation and adjustment of:

- **Cutting Platform**
- **Windrow Pickup**
- **Corn Head**
- **Feeder Conveyor**

Fig. 45 — Planetary Gear Case Location For Reverser Feature (A)

CUTTING PLATFORM ADJUSTMENTS

To operate and adjust the cutting platform properly, consider the following:

1. Ground Speed
2. Cutting Height
3. Reel Adjustments
4. Auger Adjustments

HEADER AND FEEDER HOUSE-PLUGS

Difficult field conditions or incorrect operation settings such as excess ground speed may cause plugging in the header and feeder house. Follow your operator's manual safety procedures when you manually remove plugs.

A special reverser has been developed that will remove plugs with machine power. The reverser is a planetary gearbox assembly that is controlled in the cab for operator convenience and safety. The planetary gearbox is driven by the primary countershaft. The gearbox in turn drives the header and feeder house.

When the control is actuated, the planetary gears reverse the drive to the header and feeder. The gears create slower speeds but more torque than normal forward operating conditions. The plugs are backed out of the machine.

The reverser assembly allows the operator to work at faster ground speeds without worrying about plugging. It also is a safety feature that keeps the operator in the cab during the unplugging process (Fig. 45).

GROUND SPEED

Ground speed is governed by the yield of the crop, the capacity of the combine and the skill of the operator. When the crop yields are high, ground speed should be reduced. Maximum effective ground speed, however, is affected by the capacity of the combine to thresh, separate and clean the crop. If the ground speed is too fast and overloads the combine, losses will be excessive. Even with a properly adjusted combine, the results of excessive speed are unthreshed grain or overthreshed straw. In each case, grain is lost over the straw walkers and shoe.

Usually, most operators combine at a ground speed which is easy for them to handle, and this depends a great deal on the operator's skill and condition of the crop.

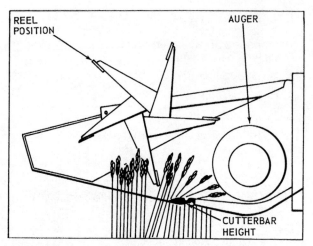

Fig. 46—Proper Cutterbar And Reel Positions In Normal Crop Conditions

CUTTING HEIGHT

Cutting height is determined by the condition of the crop. In standing grain, the cutterbar should usually be adjusted to cut low enough to get most of the grain without leaving too many low grain heads (Fig. 46). This reduces the amount of straw which would otherwise add to the load of the straw walkers and shoe. Excessive straw can cause separating and cleaning problems.

In crops which are down and tangled, lifting guards should be used to lift the crop onto the cutterbar (Fig. 47). An automatic header height control or platform float springs help maintain a close relationship of the cutterbar to the ground.

Adjust the header height during operation for variations in the height of the crop.

REEL ADJUSTMENT

The reel must be carefully adjusted to push the crop against the cutterbar and into the path of the platform auger. This requires adjustment of the reel position and reel speed.

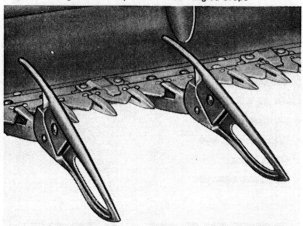

Fig. 47—Lifting Guards Help in Down and Tangled Crops

Fig. 48—Proper Reel Position In Down Crops

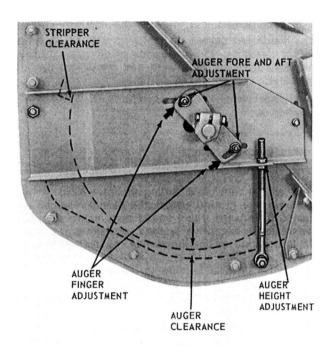

Fig. 49—Typical Auger Adjustment

Reel position in standing crops: The *reel height* is usually adjusted so that the slats of the reel touch the crop about midway between the cutoff point and the top of the crop (see Fig. 46). If the reel speed is correct, the position will cause the crop to immediately fall onto the platform when cut.

The *forward position* of the reel must be adjusted so that the reel shaft is slightly ahead of the cutterbar (see Fig. 46). If the reel is too far forward, the crop isn't pushed against the cutterbar and cut grain will fall on the ground. If the reel is too far back, the crop is pushed down too low; when it is cut, some of the heads are missed.

Reel position in down crops: A pickup reel is required. The *reel height* in down crops is low so that the fingers can lift the crop up and onto the cutterbar (Fig. 48). The fingers must be adjusted so that they pick up the crop and lift it onto the cutterbar without carrying the crop around the reel again.

The *forward position* of the reel must be about a foot (30 cm) in front of the cutterbar so that the crop is lifted before it gets to the cutterbar. Care must be used with a pickup reel so that the fingers don't come in contact with the cutterbar, which may cause the knives to break. The fingers should just clear the cutterbar. A hydraulically-powered reel fore/aft control is available. The reel position is controlled by a switch on the control panel in the operator's station.

Adjust the reel so it looks as though the reel is pulling the combine through the field. If the reel speed is too slow, the crop isn't pushed against the cutterbar; then the crop will fall on the ground because the reel doesn't push it onto the platform.

If the reel speed is too fast, the crop may be shattered by the impact of the reel. Also, the crop may be pushed down before it can be cut and the uncut grain will be left in the field.

AUGER ADJUSTMENTS

The auger must be positioned correctly and operated at the proper speed to feed the crop evenly to the feeder conveyor without bunching.

Auger position — The auger must be adjusted so that it clears the bottom of the platform by 1/4 inch to 1/2 inch (6 to 13 mm), depending on the crop (Fig. 49). If there is too much clearance, uneven feeding can result. If there is not enough clearance, the grain will be shattered and eventually fall on the ground and the straw will be chewed excessively.

The **auger stripper,** located behind the auger, must be adjusted to prevent the crop from going around the auger (Fig. 49). Poor feeding can result if too much clearance exists between the auger and stripper.

The **auger fingers** must be adjusted to provide positive feeding. In light crops, adjust the fingers to reach out and pull in the crop. In heavy crops, adjust the fingers inward. If the fingers extend too far in heavy crops, material will be carried around the auger, resulting in poor feeding. To adjust the auger on a typical combine, turn the slotted bracket down by blue arrows in Fig. 49.

Fig. 50—Auger Flight Extensions Help in Short Crops

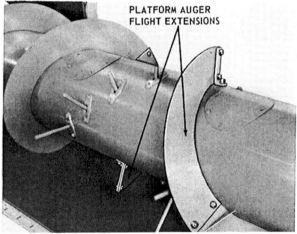

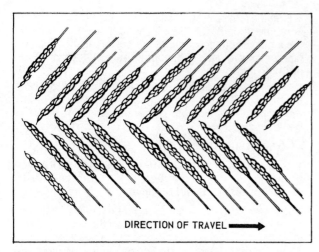

Fig. 51—Combine In The Direction That Will Pick Up Heads First

If the **auger speed** is adjustable, use the lowest speed that will give good feeding action. Fast speeds can throw the crop forward or force-feed the feeder conveyor.

If combining in short crops, auger flight extensions may be required to get the auger to feed evenly (Fig. 50). These extensions bunch the crop together before delivery to the feeder conveyor. In taller crops, the extensions should not be used. They may cause the taller crop to bunch, which will result in uneven feeding to the cylinder.

WINDROW PICKUP ADJUSTMENTS

When using a windrow pickup, keep the windrow *centered* so that the material is fed evenly to the threshing cylinder. If the heads of grain are lying in one general direction, operate the combine so that the heads are picked up first (Fig. 51). This insures better threshing and separation of the grain. Losses will also increase if the windrows are picked up butt end first.

Two adjustments must be considered when combining a windrow:

- *Speed of the pickup*
- *Position of the pickup*

Pickup Speed

Just as the reel on a cutting platform must be adjusted to the ground speed, the pickup must be adjusted to forward travel speed also. If the pickup is operated at excessive speeds, shattering will result in losses due to the impact of the pickup fingers. Fast speeds will also tear the windrow apart which will cause uneven feeding to the threshing cylinder and will result in poor threshing and separating action. Stones may also be picked up and damage the threshing cylinder.

The speed of the pickup must be adjusted to operate at a speed that makes it appear that the windrow is simply lifted up as the pickup goes underneath it. The windrow should be gathered in an unbroken, even flow (Fig. 52). Don't use ground speeds that are faster than the pickup can handle properly; bunching will occur, shatter losses will increase, and uneven feeding will result.

Pickup Position

The pickup must be positioned so that the teeth clear

Fig. 52—Gather Windrow In An Unbroken, Even Flow

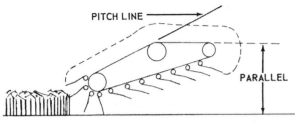

NORMAL CONDITIONS

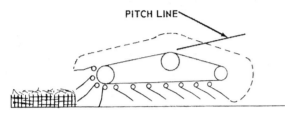

POOR WINDROW CONDITIONS

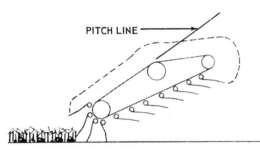

ROCKY CONDITIONS

Fig. 53—Windrow Pickup Positions

the ground yet are low enough to get all the material (Fig. 53). If the teeth are allowed to contact the ground, stones may be picked up or teeth may be bent and damaged. Also, dirt will be picked up which will make cleaning more difficult. A windrow hold-down may be used to keep the windrow down on the pickup for proper feeding.

CORN HEAD ADJUSTMENTS

Proper operation of the corn head requires careful operation and adjustment of these things:

- **Ground Speed**
- **Header Height**
- **Gatherer Points**
- **Gatherer Chains**
- **Stalk Rolls**
- **Trash Knives and Shields**
- **Snapping Plates**
- **Auger**
- **Row Spacing**

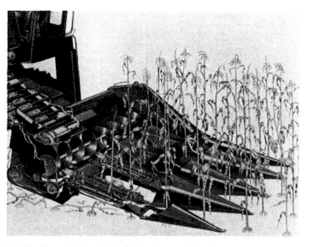

Fig. 54—Operate Header To Gather Lowest Ears Of Corn

GROUND SPEED

The proper ground speed depends on the capacity of the combine, crop condition and yield. Ground speed must not exceed the ability of the combine to harvest the crop with very little shattering and losses.

HEADER HEIGHT

The header must be operated low enough to gather the lowest hanging ears without knocking them to

Fig. 55—In Down Corn, Adjust Gatherers In Down Position

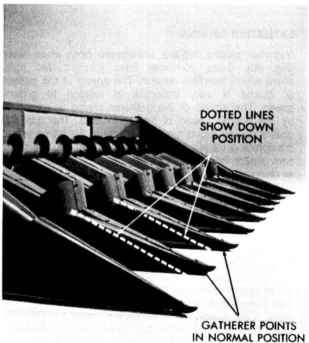

147

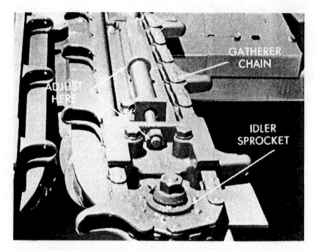

Fig. 56—Typical Gatherer Chain Adjustment For Corn Head

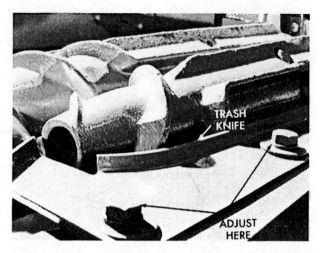

Fig. 57—Typical Adjustment of Trash Knives For Corn Head

the ground or shattering corn from the cob on impact of the header (see Fig. 54). In down and tangled corn, the header must skim the ground without picking up dirt or stones.

GATHERER POINTS

Under normal conditions, operate the gatherer points so that they follow the contour of the ground when operating in standing corn. In down corn adjust the gatherer points in a more downward position to pick up down stalks (see Fig. 55). When the corn is down and tangled, operate at a slower ground speed to reduce losses.

GATHERER CHAINS

Gatherer chains require adjustment only when worn (Fig. 56). Some combines have spring-loaded tighteners which maintain tension. The *speed* of the gatherer chains is very important in relation to ground speed. The speed of the gatherer chains is controlled by the overall speed of the corn head. The gatherer chains should move rearward at about the same speed the combine travels forward — just fast enough to guide corn stalks into the snapping rolls. Excessive gatherer chain speed causes stalks to break and throws ears to the ground. Also, the entire stalk may be pulled up and fed into the threshing unit if the chain speed is too fast. If ground speed is too fast (chain speed too slow), gatherer chains will push stalks forward and knock off ears.

STALK ROLLS

If the spacing of the stalk rolls is adjustable, adjust this distance according to the diameter of the corn stalk. The speed of the header drive controls the speed of the stalk rolls and this speed must match the forward speed of the combine to keep from losing ears.

Fig. 58—Typical Adjustment of Snapping Plates For Corn Head

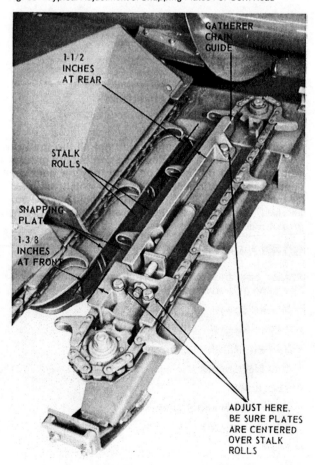

TRASH KNIVES AND SHIELDS

Trash knives and shields must be adjusted to prevent wrapping of weeds and stalks (Fig. 57). Usually, these parts are adjusted as close as possible to the stalk rolls without touching them.

SNAPPING PLATES

The snapping plates snap the ears from the stalks as the stalks are pulled down by the stalk rolls. Two snapping plates are provided for each row unit. These are adjustable to different clearances to accommodate the various sizes of stalks and ears of corn. The snapping plates are usually adjusted at about 1-3/8 inches (35 mm) apart at the front and 1-1/2 inches (38 mm) apart at the rear (Fig. 58). The wider clearance at the rear provides adequate stalk clearance so that pieces of stalk or leaves are not torn from the stalk and allowed to enter the combine. Excessive trash, such as this, will hinder the separating and cleaning action of the combine.

IMPORTANT: The snapping plates must be adjusted open as far as possible without causing shelling to minimize the amount of trash taken into the combine. The front spacing should always be approximately 1/8 inch to 3/16 inch (3.2 mm to 4.8 mm) less than the rear to maintain the proper snapping plate action.

Fig. 59—Typical Auger Adjustment For Corn Head

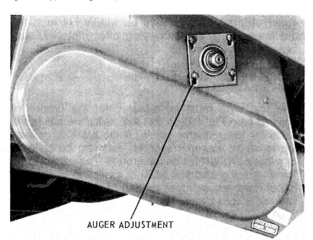

Fig. 60—Typical Adjustments For Corn Head Row Spacing

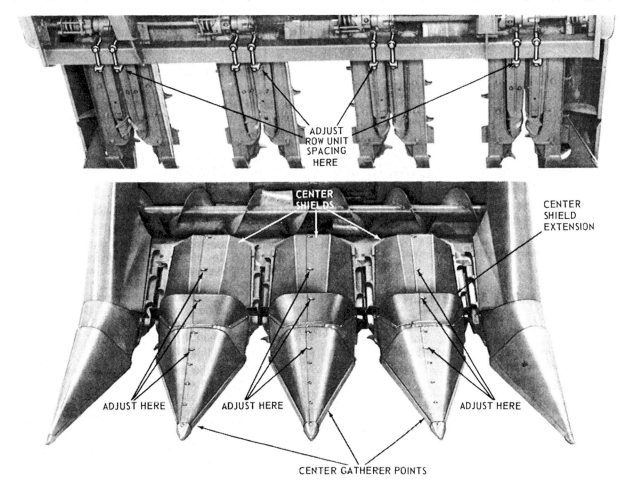

AUGER

The auger must be adjusted for the condition of the field and the size of the ears (see Fig. 59). Too much clearance can cause shelling if ears are small enough to be ground under the auger flights. Usually, the auger must be adjusted down and to the rear in normal conditions. In damp, sticky or heavy trash conditions, the auger should be adjusted slightly up and forward to move material away from the row units. The auger stripper must always be within 1/4 inch from the auger to prevent carrying over material, which can cause poor feeding.

ROW SPACING

The row units must be adjusted for the proper row spacing (see Fig. 60) so that the stalks are not pushed aside by the gatherer points. If the stalk is deflected too much by improper spacing, ears can be dislodged or missed and will fall on the ground. The difference between spacing and row width is multiplied by the number of rows, so, improper spacing becomes very serious with wider corn heads.

Fig. 62—Adjust Ground Speed To Crop And Ground Conditions

GROUND SPEED

Always adjust ground speed to combine capacity and ground and crop conditions (Fig. 62). Excessive speed increases shatter losses and chances of plugging the row-crop head. If speed is too slow, combine capacity and time are wasted. Also, gathering chains may pull plants up by the roots causing excessive shattering and poor cutting action.

GATHERING BELT SPEED

Gathering belts should operate at approximately the same speed as the combine to prevent excessive shattering of seeds, plugging of row units or pulling plants out of the ground before they are cut off. Gathering belt speed may be adjusted by changing drive chain sprockets or adjusting the variable speed feeder house drive, if used.

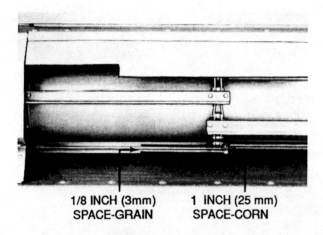

1/8 INCH (3mm) SPACE-GRAIN   1 INCH (25 mm) SPACE-CORN

Fig. 61—Typical Feeder Conveyor Chain Adjustment

ROW-CROP HEAD ADJUSTMENTS

The row-crop head performs best when all components are properly adjusted and function together. This requires proper adjustment or control of the following factors:

- Ground Speed
- Gathering Belt Speed
- Gatherer Point Clearance and Spacing
- Operating Height

GATHERER POINT CLEARANCE AND SPACING

Gatherer points must be set high enough to prevent them from scooping dirt into the gathering belts, but low enough to slide under low-growing or lodged plants. If soybeans are standing well, set points to clear the ground by about 2 inches (50 mm). However, if beans are lodged, lower gatherer points to about ½ to 1 inch (13 to 25 mm) from the ground. On wide row-crop heads (6 or 8 rows), outer gatherer points may be set slightly higher than the other points for smoother operation over uneven terrain.

Always operate the row-crop head on the same rows as they were planted to avoid cutting odd or guess rows which may not properly align with row unit spacing. Also be sure that row unit spacing on the row-crop head matches the spacing between crop rows to avoid excessive shattering and crop losses. Row units can be shifted on the head to match row spacings within certain limitations but cannot be set to match all possible row widths.

Fig. 63—Float Spring Tension Adjustment

Fig. 64—Perforated Feeder House Bottom Insert

OPERATING HEIGHT

Each row unit on the row-crop head is designed to float independently for close cutting of low-growing crops. They also may be locked rigidly in place for operation in taller crops such as sorghum or sunflowers. To permit free flotation of row units, be sure there is adequate clearance between the gatherer shields over each row unit.

Float springs on each row unit must also be adjusted so that units float freely over uneven terrain, but drop quickly to lower levels when necessary (Fig. 63). Skid shoes on each row unit may be adjusted up for closer cutting, or down to increase cutting height in low-growing crops such as soybeans. Automatic header height control automatically raises or lowers the entire row-crop head whenever changes in ground elevation cause any row unit to move up or down more than the float range provided between units. For instance, if the float range is 6 inches (15.25 cm) and one row unit passes over an object which is 7 inches (18 cm) high, the entire row-crop head will be automatically raised 1 inch (2.5 cm) until the obstruction is passed, and then lowered to the proper height above the ground.

FEEDER CONVEYOR ADJUSTMENTS

The feeder conveyor chain must operate properly to feed the material smoothly and evenly to the threshing cylinder. The chain must be tight enough so the slats don't slap the bottom of the feeder house. On combines which have adjustable conveyors, adjust the chain clearance for the size of crop. For grain crops the space beneath the slats should be about 1/8 inch (3 mm) and for corn about 1 inch (25 mm) (Fig. 61).

When combining large seed crops in dirty or weedy conditions, a perforated feeder house bottom insert should be used (Fig. 64). This allows dirt and weed seed to fall through the holes resulting in cleaner grain.

OPERATING AND ADJUSTING THRESHING UNITS

The threshing section of the combine is the "heart" of the harvesting function. Many combine problems can occur if the operator does not know why or how to make the proper adjustments.

GENERAL RULES

The cylinder or rotor and concave must be adjusted as follows:

- In small grain and seeds, all grain must be removed from the stalk without damaging the grain, breaking the straw or causing excess chaff.
- In corn, all kernels must be removed from the cob without damaging the kernels or breaking many of the cobs.

The general rule of thumb is:

- **Small grain,** *high* threshing speeds and *narrow* concave spacing
- **Large grain or seeds,** *low* threshing speeds and *wide* concave spacing.

THRESHING ACTION

Remember: Usually, the higher the cylinder or rotor speed the narrower the concave spacing. This provides the best threshing action; the opposite will decrease threshing action.

Here are the results of overthreshing, underthreshing, and proper threshing:

Overthreshing
- *Cracked grain*
- *Broken and chewed straw which overloads the shoe*
- *Grain losses over the shoe*

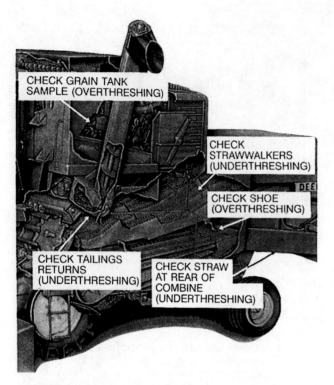

Fig. 65—Check Threshing Results In These Areas

Underthreshing
- *Unthreshed heads and excessive tailings*
- *Overloaded straw walkers*
- *Grain losses over the straw walkers*

Proper Threshing
- *No cracked grain*
- *No unthreshed heads or excessive tailings*
- *No broken or chewed straw*
- *No grain losses over the straw walkers and shoe*

Fig. 66—Check The Grain Tank For Cracked Kernels

HOW TO DETERMINE THRESHING ACTION

If the operator is to correct improper threshing action, he must be able to recognize it and determine what is causing it. To do this he must examine the straw and grain discharged from the rear of the combine, check the grain tank sample, the tailings returns, straw walkers and shoe (Fig. 65).

Straw discharged from the rear of the combine should not be broken or chewed and very few kernels of grain should still be attached to the stalk. In corn, the cobs must not be broken excessively and few if any kernels should remain on the cob.

The grain tank sample should have very few cracked kernels (Fig. 66). If many cracked kernels are found, the problem may be excessive tailings returned to the cylinder for rethreshing rather than fast cylinder speeds or narrow concave spacing.

Tailings returns should be very few. Each paddle of the elevator should contain not much more than a tablespoon of tailings or less and the sample must contain mostly unthreshed heads rather than loose kernels or straw and chaff.

Straw walkers will be overloaded if *underthreshing* is occurring at high speeds. Grain cannot be separated from the thick masses of straw on the walkers. Thus, grain is carried out the rear of the machine with the straw.

The **cleaning shoe** will be overloaded with excessively chewed and broken straw when *overthreshing* is occurring. Kernels cannot be separated from the mass of straw and chaff on the chaffer. Thus, grain is carried out the rear of the machine with the straw.

HOW TO CORRECT THRESHING PROBLEMS

After determining that the problems may be improper threshing rather than excessive tailings or a poorly adjusted shoe, use one of the following procedures:

Combine operating speed may be too fast or too slow. If engine or primary countershaft speed is more than ±10 rpm off the recommended speed, have the engine adjusted to the proper speed. Adjust governor on gasoline engine.

Overthreshing may be reduced by slowing cylinder or rotor speed by five percent at a time (Fig. 67). Check the results of these changes before making additional changes. Try opening the concave spacing slightly if reducing threshing speed 10 percent doesn't help. If overthreshing cannot be corrected by these measures, try slowing the ground speed; too much material at too fast a feed rate may cause overthreshing.

Underthreshing may be caused by too slow a cylinder or rotor speed and wide concave spacing (Fig. 68). Try increasing the threshing speed by five percent. If this doesn't correct the problem, narrow the concave spacing slightly. Check the results of these changes frequently. If after trying a few of these adjustments threshing action still hasn't improved, increase ground speed by about half a gear range. Check the results and make additional adjustments as necessary.

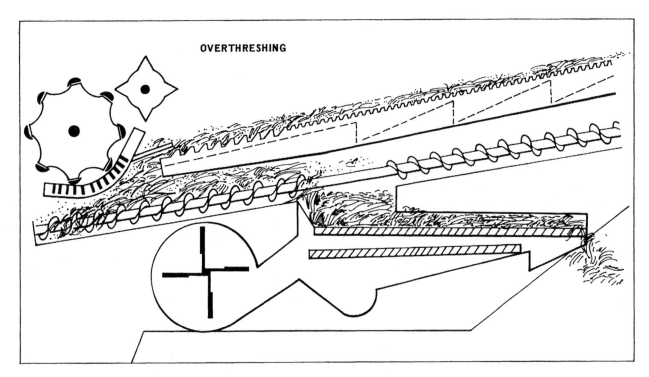

Fig. 67—Overthreshing Chews and Breaks The Straw, Which Overloads The Shoe

OPERATING AND ADJUSTING CLEANING UNITS

The cleaning fan, chaffer and sieve must be operated and adjusted in conjunction with each other to get the best cleaning action. Fine chaff and straw must be removed with as much air as possible without blowing kernels out of the combine.

Fig. 68—Underthreshing Overloads The Walkers

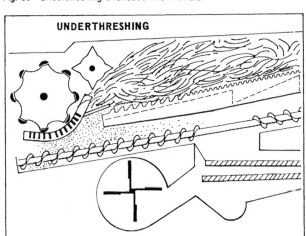

Fig. 69—Adjust The Fan To Lowest Suggested Speed

ADJUSTING THE FAN SPEED

Before adjusting the fan speed, open the chaffer and sieve to the maximum recommended openings for the crop being harvested. Then start with the lowest fan speed suggested and gradually increase the speed without blowing kernels out of the combine or into the tailings return. Check the results carefully. After reaching the maximum acceptable fan speed, continue to make minor adjustments to the fan speed and to chaffer and sieve openings until best results are achieved.

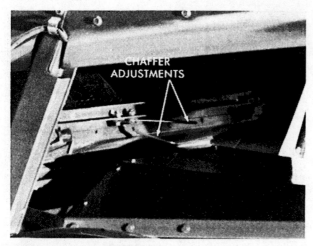

Fig. 70—Open Chaffer And Sieve To Maximum Recommended Openings

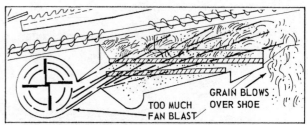

Fig. 72—Too Much Fan Blast Will Blow Grain Over The Shoe

ADJUSTING THE CHAFFER

Open or close the chaffer just enough so that the grain falls through before passing the length of the chaffer. If the chaffer is opened too wide, it may overload the sieve with chaff and straw and increase the tailings returns. If the chaffer is not opened wide enough, excess grain will be moved to the tailings return and some will be lost out the rear of the combine. Most combines have a chaffer extension which should be opened slightly more than the chaffer and raised a little to allow tailings to drop through easily.

Grain losses in the cleaning unit may be caused by:

1. Too little fan blast or too narrow chaffer openings which causes a layer of straw and chaff to carry over the grain (Fig. 71).

2. Too much fan blast which will blow the grain over the shoe (Fig. 72).

It is important to know which of the above reasons is causing the shoe losses so that you can make the proper adjustment to the area which is causing the problem. *In item one above,* the problem may be caused by overthreshing as well as too little air from the fan. Check the amount and condition of the straw; if the straw appears excessively broken and chewed, the cylinder and concave must be adjusted to reduce overthreshing. If the straw is whole and unbroken, then more air is needed to suspend the straw and chaff so that the grain can drop through to the sieve, or the chaffer needs to be opened slightly. Be sure to check the results before going too far.

If too much fan blast is the problem, then there will be very little chaff and straw on the shoe, which indicates too much air is being used. Reduce the fan speed and check the results.

ADJUSTING THE SIEVE

The sieve does the final job of cleaning. It must be opened far enough to allow the kernels to fall through easily, but not so far that chaff and straw are allowed to drop through.

If the sieve is closed too far, the grain will move to the tailings return and overthreshing will occur because of excessive tailings returns. This may also cause excessive kernel cracking.

To adjust the sieve, open it until too much foreign material appears in the grain tank and then close the sieve slightly until the grain tank sample becomes acceptable (see Fig. 70). If the combine has a lower position for the front of the sieve, use this position because it will retard the movement of the material and allow better cleaning.

DETERMINING GRAIN LOSSES

After adjusting the harvesting units to get the best job, the operator must also determine what *grain losses* are occurring. The few minutes spent in determining machine losses and making the proper adjustments are well worth the time.

Suppose you are harvesting a 20-acre (8.1 ha) corn field with an average yield of 100 bushels per acre (5 tonnes/ha). After checking the machine losses, you find you are losing six bushels an acre (300 kg/ha) in the harvesting process. However, if it is properly adjusted, the combine should lose not more than three bushels per acre (150 kg/ha) in this field. *By properly adjusting the machine, you can save three bushels an acre (150 kg/ha).* Figuring corn at $2.50 per bushel ($98/tonne) in the fifteen or twenty minutes it takes to measure the loss and make the corrective adjustments, you can save $150.00 for the 20 acre (8.1 ha) field!

Fig. 71—Too Little Fan Blast Or Too Narrow Chaffer Openings Cause Layer Of Chaff And Straw To Carry Grain Out Of Combine

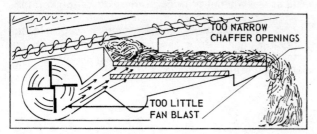

Fig. 73—Check Losses Occasionally To Determine Combine Efficiency

Grain losses result from improper adjustment of the combine. Even if the operator has adjusted the machine to cut, thresh, separate and clean the grain acceptably, he should check the combine occasionally to make sure it is continuing to operate properly.

If the grain losses are not acceptable, the operator must reduce them by adjusting the components which are causing the costly losses.

To reduce losses, the operator must know:

1. *The source of losses*
2. *How to measure losses*

SOURCES OF GRAIN LOSSES

Unless the operator knows the source of his grain losses, he cannot reduce them. Some losses are due to improper operation and others are caused by improper adjustment.

Here are the sources of losses the operator must identify:

- **Preharvest losses**
- **Header losses**
- **Threshing losses**
- **Straw walker losses**
- **Shoe losses**
- **Leakage losses**

PREHARVEST LOSSES

Preharvest losses are those which occurred in the field before combining. Such losses show up as grain on the ground as a result of wind shatter, lodging, down crop or weather conditions as shown in Fig. 75. Corn and soybeans are two common crops which may have large preharvest losses.

Fig. 74—Typical Sources of Losses

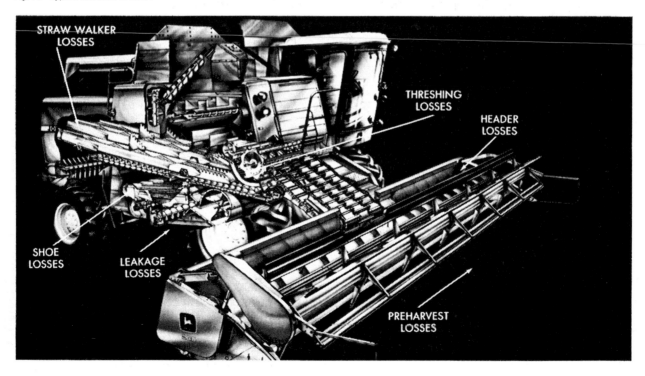

Fig. 75—Weather Conditions May Cause Great Preharvest Losses

HEADER LOSSES

These losses occur when the header is operated improperly or when the crop tends to shatter easily. Each type of header has operating characteristics which can cause losses. Losses caused by faulty adjustment and operation of cutting platforms, corn heads and windrow pickup units are discussed below.

Cutting Platform Losses

Usual causes of cutting platform losses are:

1. *Grain heads missed by cutterbar*
2. *Grain shattered by action of knife*
3. *Grain missed because of improper reel speed*
4. *Grain shattered by too-fast a reel speed*
5. *Grain thrown over in front of the reel by too-low a reel height*
6. *Grain shattered by too-fast a ground speed*
7. *Grain uncut by damaged knife or guard.*

Corn Head Losses

Usual causes of corn head losses are:

1. *Ears missed by gatherers*
2. *Kernels shattered from cobs by impact with header*
3. *Ears missed because of improper gatherer chain speed*
4. *Ears deflected to ground by too-fast ground speed*

Windrow Pickup Losses

Usual causes of windrow pickup losses are:

1. *Grain shattered by too-fast a pickup speed*
2. *Grain missed because of improper pickup height*
3. *Grain shattered by too-fast a ground speed*
4. *Grain shattered by too-slow a pickup speed*

THRESHING LOSSES

Threshing losses are caused by:

1. *Unthreshed grain carried over straw walkers*
2. *Cracked grain due to overthreshing*
3. *Cracked grain due to excessive tailings*

STRAW WALKER OR SEPARATOR LOSSES

Straw walker or separator losses are usually caused by feeding too much material over them at slow cylinder speeds and wide concave spacing when the combine is operating at excessive ground speeds. Too much material prevents the grain from falling through the walkers and onto the cleaning shoe.

156

Fig. 76—Keep Doors And Covers Tightly Closed To Prevent Leakage Losses

SHOE LOSSES

Shoe losses may be caused by:

1. *Too much air from fan*
2. *Too much material on chaffer*
3. *Improperly adjusted chaffer and sieve*

TOO MUCH AIR from excessive fan speeds will blow chaff and grain over the shoe and out the rear of the combine.

TOO MUCH MATERIAL deposited on the chaffer from overthreshing prevents the grain from falling through to the sieve and the fan blast cannot blow the straw and chaff free of the chaffer. Thus the grain rides out the rear of the combine with the straw and chaff.

AN IMPROPERLY ADJUSTED CHAFFER AND SIEVE will not allow grain to fall through if openings are not wide enough. Grain either rides out the rear of the combine on the chaffer, or it is delivered to the tailings return system which delivers the grain to the cylinder for rethreshing. Too many tailings can cause excessive kernel crackage.

LEAKAGE LOSSES

Leakage losses can occur almost anywhere on the combine. To guard against leakage, inspect the combine to see that all inspection doors, cleaning doors and drainage doors are in the proper position and closed securely. Also check for torn seals, damaged sheet metal or holes.

All leaks must be repaired before accurate measurement of losses can be made. Otherwise, it is difficult to determine where losses are occurring.

HOW TO MEASURE LOSSES

We will now discuss these losses as they pertain to harvesting *small grain* and then *corn*.

HOW TO DETERMINE LOSSES IN SMALL GRAINS AND SOYBEANS

The following losses can be measured to determine where losses are occurring:

1. **Preharvest losses**
2. **Header Losses**
3. **Threshing losses**
4. **Straw walker (separator) and shoe losses**

When checking small-grain and soybean losses, always use a typical harvest area, well in from the edges of the field. It is best to follow sequential steps to avoid mistakes while determining losses.

After each explanation of the checking areas, a typical example of a loss problem is given. This example is used only to make the explanation clearer and should not be used as a guide in field operation.

This procedure may be used to determine losses in wheat, oats, barley, soybeans, etc. Follow the procedures outlined below to find where losses must be reduced.

First determine preharvest losses so that you can compare this with the machine losses. The difference in these two figures will indicate whether or not you are doing a good job of combining and whether or not adjustments are necessary.

PREHARVEST AND HEADER LOSS CHART — SMALL GRAINS AND SOYBEANS		
(Approximate Number Of Kernels Per Square Foot To Equal One Bushel Per Acre, or Per 1/10 m² To Equal 50 kg/ha)		
Crop	Kernels Per Sq. Ft.	Kernels Per 1/10 m²
Barley	13-15	13-15
Beans-Red Kidney	1-2	1-2
Beans-White Pea	3-4	3-4
Oats	10-12	15-18
Rice	29-31	31-33
Rye	21-24	18-21
Sorghum	19-22	16-19
Soybeans	4-5	3-4
Wheat	18-20	14-16

HOW TO DETERMINE PREHARVEST LOSSES

To determine preharvest losses, select a typical unharvested area of the field well in from the edges (Fig. 76). Place a frame 12 inches square (One-square-foot, or make it 31.6 cm square — equal to 1/10 m²) in the standing crop. Count all the kernels laying on the ground within the frame. Refer to the *Preharvest and Header Loss Chart* (above) to determine how much grain is already on the ground. Make several random samples and average them to find average grain lost per acre (hectare).

Fig. 77—Losses In Small Grain Must Be Measured To Determine Combine Efficiency

a distance equal to the length of the machine and stop the combine. This will allow you to check all the loss points without starting and stopping the combine several times.

HOW TO DETERMINE HEADER LOSSES

To determine header losses, after backing the length of the machine, place the one-square-foot measuring frame on the ground in front of the combine within the harvested area (see Fig. 78). Count the number of kernels found in the frame. Check several other sample areas and average the kernel count. Finally, subtract the number of kernels found in the preharvest loss check. Use the *Preharvest and Header Loss Chart* (page 155) to determine the loss in bushels per acre (kg/ha).

For example, if the combine has a 14-foot (4.3 m) cutting platform and is harvesting wheat, suppose you find 39 kernels of wheat within the frame. Preharvest losses were 20 kernels. Subtract the preharvest loss and you are left with 19 kernels. By checking the *Preharvest and Header Loss Chart,* we find the header loss is one bushel per acre (67 kg/ha).

Typical header losses for small grain crops, when the combine is adjusted and operated correctly, may vary from 1/2 to 2 percent of the average yield. For soybeans the range is from 10 to 12 percent of the average yield.

If losses are not acceptable, consult the troubleshooting chart on *Field Problems* (page 165) to determine possible remedies for header losses.

For example, if you are harvesting wheat and find 20 kernels within the frame, the preharvest loss is one bushel per acre (67 kg/ha).

HOW TO DETERMINE MACHINE LOSSES

When checking machine losses, do not use any straw spreading device, such as a straw chopper or straw spreader, because the loss count will be inaccurate (Fig. 77). Harvest a typical area. Allow the machine to clear itself of material and then back the combine

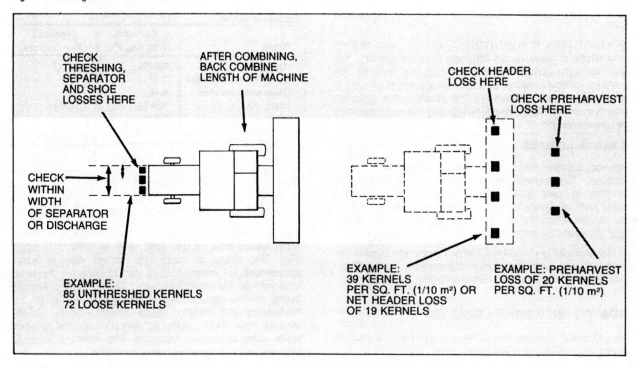

Fig. 78—Checking Small Grain Losses

MACHINE LOSS CHART FOR SMALL GRAIN*											
(Approximate Number Of Kernels Per Square Foot To Equal One Bushel Per Acre)											
Crop	Separator Width (in.)	Cutting Width (Ft.)									
		10	13	14	15	16	18	20	22	24	30
Barley	38	55	71	76	82	87	98	109	—	—	—
	44	47	61	66	71	76	85	95	104	—	—
	55	—	49	53	57	61	68	76	83	91	113
	60	—	45	48	52	55	62	69	76	83	104
	65	—	—	—	48	51	57	64	70	77	96
Beans-Red Kidney	38	4	5	6	6	7	8	9	—	—	—
	44	4	5	5	6	6	7	7	8	—	—
	55	—	4	4	4	5	5	6	6	7	9
	60	—	4	4	4	4	5	5	6	6	8
	65	—	—	—	4	4	4	5	5	6	7
Beans-White Pea	38	7	9	10	10	11	12	14	—	—	—
	44	6	8	8	9	10	11	12	13	—	—
	55	—	6	7	7	8	9	10	11	12	14
	60	—	6	6	7	7	8	9	10	11	13
	65	—	—	—	6	6	7	8	9	10	12
Oats	38	29	38	41	44	47	53	59	—	—	—
	44	25	33	36	38	41	46	51	56	—	—
	55	—	26	28	30	32	37	41	45	49	61
	60	—	24	26	28	30	33	37	41	45	56
	65	—	—	—	26	27	31	34	38	41	52
Rice	38	53	69	75	80	85	96	107	—	—	—
	44	46	60	65	69	74	83	92	101	—	—
	55	—	48	52	55	59	66	74	81	88	111
	60	—	44	47	51	54	61	68	74	81	101
	65	—	—	—	47	50	56	62	69	75	94
Rye	38	116	151	163	174	186	209	232	—	—	—
	44	100	130	141	151	161	181	201	221	—	—
	55	—	104	112	120	128	145	161	177	193	241
	60	—	96	103	110	118	132	147	162	177	221
	65	—	—	—	102	109	122	136	149	163	204
Sorghum	38	61	80	86	92	98	110	123	—	—	—
	44	53	69	74	79	85	95	106	116	—	—
	55	—	55	59	63	68	76	85	93	102	127
	60	—	50	54	58	62	70	78	85	93	116
	65	—	—	—	54	57	64	72	79	86	107
Soybeans	38	11	15	16	17	18	20	23	—	—	—
	44	10	13	14	15	16	18	20	22	—	—
	55	—	10	11	12	13	14	16	17	19	24
	60	—	9	10	11	12	13	14	16	17	22
	65	—	—	—	10	11	12	13	15	16	20
Wheat	38	61	79	85	91	97	109	121	—	—	—
	44	52	68	73	79	84	94	105	115	—	—
	55	—	54	59	63	67	75	84	92	101	126
	60	—	50	54	58	61	69	77	84	92	115
	65	—	—	—	53	57	64	71	78	85	106

*See page 212 which gives comparable chart with dimensions in metric units

HOW TO DETERMINE THRESHING LOSSES

To determine threshing unit loss, after backing the length of the machine, check the ground in a few places directly behind the separator or machine discharge, using the one-square-foot (1/10 m²) frame (see Fig. 78). Count all the kernels remaining on partially threshed heads or pods. Do not include kernels lying loose on the ground. Then check the *Machine Loss Chart For Small Grain* (above) to determine the loss per acre. Check *Machine Loss Chart For Small Grain* on page 212 of Appendix to determine grain loss per hectare.

Fig. 79—Do Not Use A Spreading Device When Checking Losses

For example, if a combine with a 14-foot (4.3-meter) cutting platform and 38-inch (100-cm) separator were being used to harvest wheat and 85 kernels were found on partially threshed heads, the loss would be one bushel per acre (59 kg/ha).

Losses On Combines With Different Separator Width

The chart does not list all possible separator or discharge widths. For combines with discharge widths different from those listed, the procedure for determining threshing losses is basically the same.

Again, place the frame on the ground in a few places directly behind the machine discharge and count the number of kernels remaining on partially threshed heads or pods within the frame. Then find the percentage by which the discharge width varies from one of the discharge widths listed in the chart. Multiply the number of kernels counted by this percentage.

For example, suppose the discharge or separator width on a combine is 24 inches (60 cm). The combine is using an 18-foot (5.5 meter) header. Place the frame within this 24-inch (60-cm) wide path and count the number of kernels on partially threshed heads or pods.

The 24-inch (60-cm) discharge width is 63 percent as large as the 38-inch (100-cm) separator listed in the chart. If 155 kernels were counted within the frame, mutiply: 155 × .63 = 98. Compare this to the chart (38 inch separator, 18 ft. head) and find that this figure equals one bushel per acre loss. Check *Machine Loss Chart For Small Grain* on page 212 of Appendix to determine grain loss per hectare.

Typical threshing unit loss ranges from 1/2 to 1 percent of the average yield. Acceptable losses are largely a matter of operator preference. If losses are not acceptable, consult the chart on *Field Problems* (page 165) to determine possible remedies for threshing unit loss.

HOW TO DETERMINE SEPARATOR AND CLEANING SHOE LOSSES

To determine separator and shoe losses, after backing the length of the machine, place the one-square-foot measuring frame on the ground directly behind the separator or machine discharge (see Fig. 78). Then count the kernels lying loose within the frame. Do not include kernels on partially threshed heads. On combines with separator widths different than those listed in the chart, multiply the number of kernels found lying loose within the frame by the percentage the width varies.

Then subtract the number of kernels found in the header-loss check and the preharvest-loss check. The remaining figure will be the number of kernels lost over the separator and shoe. Check the *Machine Loss Chart For Small Grain* (page 159) to find the loss per acre (hectare).

Fig. 80—Conditions Such As This Increase Combining Losses

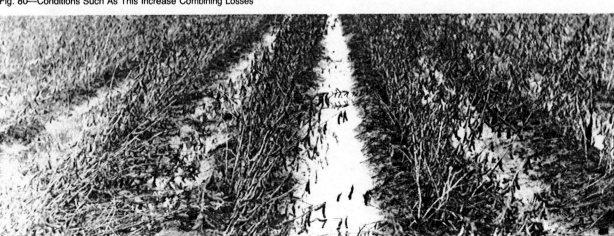

Fig. 81—Losses In Corn Must Be Measured To Determine Combine Efficiency

For example, if a combine with a 14-foot (4.3-meter) cutting platform and a 38-inch (100-cm) separator is being used to combine wheat, suppose you find 72 loose kernels within the frame. On a machine with a 24-inch (60 cm) separator or discharge, 115 kernels were counted. Multiply 115 × .63 = 72 kernels. Remember, 20 of the kernels are from preharvest loss and 19 of the kernels are from header loss. By subtracting these counts (total 39) from 72, the example shows 33 kernels are being lost over the straw walkers (or separator) and shoe. Checking the *Machine Loss Chart For Small Grain* (page 159) we find the loss is about 1/3 bushel per acre (18 kg/ha).

Typical separator and shoe losses should be less than one percent of the average yield. Acceptable losses are largely a matter of operator preference.

If losses are not acceptable, consult the chart on *Field Problems* (page 167) to determine possible remedies for separator and shoe loss.

To check for shoe losses, have someone walk beside the machine as it is combining and hold a scoop shovel behind and below the rear of the shoe to catch the material as it falls off. There should be very few kernels in the sample. If several kernels are found in the shoe sample, adjust the shoe.

ACCEPTABLE LOSSES IN SMALL GRAINS

What are acceptable losses? They may vary with the operator. Some operators aren't concerned with machine losses of five percent and others don't want any. *Generally, acceptable crop losses range from three to five percent of the yield.*

HOW TO DETERMINE LOSSES IN CORN

While procedures used to check losses in corn are similar to those used in small grain, some items are different. The following points should be checked to determine corn losses:

1. **Preharvest ear losses**
2. **Corn head ear losses**
3. **Corn head kernel losses**
4. **Threshing unit losses**
5. **Separator and shoe losses**
6. **Leakage losses**

When checking corn losses, always use a typical harvest area, well in from the edges of the field. It is best to follow steps in sequence to avoid mistakes while determining losses.

After each explanation of inspection areas, a typical example of a loss problem is given. This example is used only to make the explanation clearer and should not be used as a guide in field operation.

It is best to determine preharvest losses first so that these losses can be compared with machine losses.

The difference in these two figures will indicate whether you are doing a good job combining and if adjustments are necessary.

HOW TO DETERMINE PREHARVEST LOSSES

To determine preharvest ear losses, select a typical unharvested area of the field, well in from the edges (Fig. 82). Place a marker on one of the rows and measure the necessary distance for 1/100 acre (1/200 ha) according to the chart on page 162. Use the appropriate figures for the row spacing of the corn and the number of row units on the corn head.

Gather all ears found on the ground in this area. Count each 3/4-pound (1/3 kg) ear or equivalent smaller ears as one bushel lost per acre (50 kg/ha of shelled corn). For example, if two 3/4 pound (1/3 kg) ears are found in the area, preharvest loss is two bushels per acre (100 kg/ha).

HOW TO DETERMINE CORN HEAD EAR LOSSES

After operating the combine for some distance, check the rows harvested behind the combine (see Fig. 82). Then count the number of 3/4-pound (1/3 kg) ears picked up in 1/100 acre (1/200 ha). Subtract the preharvest loss figure from this amount to determine corn head ear loss.

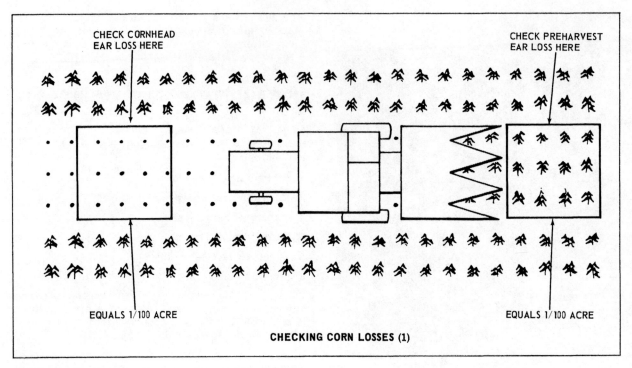

Fig. 82—Checking Preharvest Ear Loss and Corn Head Ear Loss

For example, if four 3/4-pound (340 g) ears are found in 1/100 acre (1/200 ha) and preharvest loss was two 3/4-pound (340 g) ears (two bushels per acre or 100 kg/ha), corn head ear loss equals two bushels per acre (100 kg/ha).

Less than one percent loss can be expected when the machine is adjusted and operated correctly. A one to two bushel (60 to 120 kg/ha) loss may be considered acceptable, but this depends on operator judgment. If losses are not acceptable, consult the troubleshooting chart on *Field Problems* (page 167) to determine possible remedies for corn head ear losses.

DETERMINING MACHINE KERNEL LOSSES

When checking kernel losses, first harvest a typical area. Allow the machine to clear itself of material and then back the combine a distance equal to the length of the machine and stop the combine. This will allow you to check all loss points without starting and stopping the combine several times.

If machine losses are not acceptable, check the individual units of the combine as described in the following paragraphs to determine where the losses are occurring.

HOW TO DETERMINE CORN HEAD KERNEL LOSSES

To determine corn head kernel losses, after backing the length of the machine, use a measuring frame to enclose 10 square feet (1 m²) of the sample area directly in front of the corn head (Fig. 83). Consult the *Loss Measuring Frame Dimension Chart* (on page 163) to determine the correct inner dimensions of the frame. Then count the number of kernels within the frame. Twenty kernels per 10

	ROW LENGTH IN FEET PER 1/100 ACRE*						
	Distance Per 1/100 Acre For:						
Row Width-Inches	One Row	Two Rows	Three Rows	Four Rows	Six Rows	Eight Rows	Twelve Rows
20	262	131	87.3	65.5	43.6	32.7	21.8
28	187	93.5	61.3	46.7	31.1	23.4	15.6
30	174	87	58	43.6	29	21.8	14.5
36	145	72.5	48.3	36.2	24.2	18.1	12.1
38	138	69	46	34.5	23	17.2	11.5
40	131	65.5	43.6	32.7	21.8	16.4	10.9
*See page 162 for comparable chart with dimensions in metric units.							

LOSS-MEASURING FRAME DIMENSION CHART FOR CORN			
Frame Size for 10 Square Feet (1 m²)			
Row Width		Row Length	
Inches	(cm)	Inches	(cm)
20	(50) by	72	(200)
28	(70) by	51.5	(143)
30	(75) by	48	(133)
36	(90) by	40	(111)
38	(95) by	38	(105)
40	(100) by	36	(100)

MACHINE-KERNEL LOSS CHART FOR CORN*						
(Approximate Kernels Per 10 Square Feet To Equal One Bushel Per Acre)						
Corn Head Size	Row Spacing (Inches)	Separator Width (Inches)				
		38	44	55	60	65
3-Row	36	52	45	—	—	—
	38	55	48	—	—	—
	40	58	50	—	—	—
4-Row	28	54	47	—	—	—
	30	58	50	—	—	—
	32	62	54	—	—	—
	36	70	60	48	44	—
	38	74	64	51	47	—
	40	77	67	54	49	—
5-Row	28	—	59	47	—	—
	30	—	63	50	—	—
	36	—	75	60	55	51
	38	—	79	64	58	54
	40	—	84	67	61	57
6-Row	28	—	70	56	52	48
	30	—	75	60	55	51
	36	—	90	72	66	61
	38	—	95	76	70	65
	40	—	100	80	74	68
8-Row	28	—	94	75	67	63
	30	—	100	80	74	68
	36	—	—	96	88	82
	38	—	—	102	93	86
	40	—	—	107	98	91
12-Row	28	—	—	112	103	95
	30	—	—	120	110	102

*See page 211 for comparable chart with dimensions in metric units.

square feet equals one bushel per acre (16 kernels per square meter equals 50 kg/ha). For example, if 10 kernels are found in the 10-square-foot (1 m²) area, the loss is 1/2 bushel per acre (25 kg/ha). Expected normal loss is less than one percent of the average yield when the combine is adjusted and operated properly.

If losses are not acceptable, consult the troubleshooting chart on *Field Problems* (page 167) to determine possible remedies for corn head kernel loss.

HOW TO DETERMINE THRESHING UNIT LOSSES

To determine threshing unit losses, check the ground in a few places directly behind the separator or machine discharge (see Fig. 83). Use the 10-square-foot (1 m²) frame and count all kernels remaining on the cobs. Do not include kernels lying loose on the ground. Combines may have separator widths different than those listed in the chart in column 2. For these combines, use a 10-square-foot frame with the width dimension no greater than the width of the separator or machine discharge. Again count the kernels remaining on the cobs within the frame. Multiply this number by the percentage the separator or discharge width varies from one of the widths listed. Then check the *Machine Kernel Loss Chart* (in column 2) to determine the loss in bushels per acre. For example, if a combine with a three-row corn head and 38-inch separator were being used on 40-inch rows and 29 kernels were found on threshed cobs in a 10-square-foot area, threshing unit loss would be one-half bushel per acre. If this combine had only a 24-inch (60-cm) separator, 46 kernels would have been counted on the threshed cobs. This separator is 63 percent as wide as the 38-inch (100-cm) separator. Therefore, multiply 46 × .63 = 29 kernels to obtain the equivalent threshing unit loss.

Less than one percent of the total yield may be lost when the combine is operating properly. If losses are not acceptable to the operator, consult the troubleshooting chart on *Field Problems* (page 167) to determine possible remedies for threshing unit losses.

HOW TO DETERMINE SEPARATOR AND SHOE LOSSES

To determine separator and shoe losses, after backing the machine, use a 10-square-foot (1 m²) frame to check several areas directly behind the separator (see Fig. 83). Consult the table shown for proper inside dimensions of the frame (one dimension should be no wider than the width of the separator or machine discharge). Then count the number of loose kernels lying on the ground. Do not include kernels remaining on threshed cobs (threshing unit losses). For combines with separator widths different than those listed in the chart, multiply the number of kernels counted by the percentage the width varies.

Then subtract the number of kernels found under corn head kernel losses. Use the *Machine Kernel Loss Chart* (above) to determine total losses per acre from the remaining kernel count.

By checking the *Machine Loss Chart* it is determined that one bushel per acre is being lost on the separator and shoe.

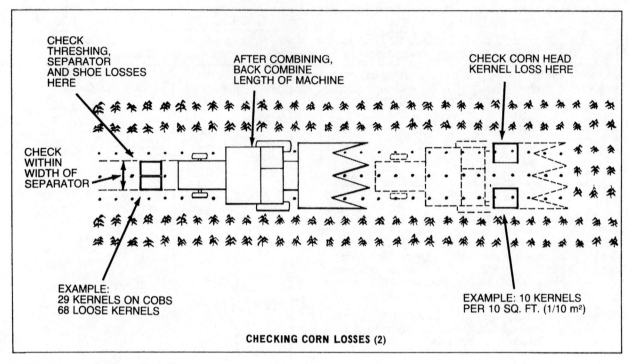

Fig. 83—Checking Kernel Losses in Corn

Less than one percent of the total yield may be lost on the straw walkers and shoe when the combine is adjusted and operated properly. If the losses are unacceptable, consult the troubleshooting chart on *Field Problems* (page 167) to determine possible remedies for straw walker and shoe losses.

To determine shoe losses, have someone walk beside the machine as it is combining and hold a scoop shovel behind and below the rear of the shoe to catch the material as it falls off. There should be very few kernels in the sample. If several kernels are found in the shoe sample, the shoe must be adjusted.

ACCEPTABLE LOSSES FOR CORN

What are acceptable losses? They may vary with the operator. Some operators aren't concerned with total machine losses of five percent and others don't want any. *Generally, total acceptable corn losses range from three to five percent of the yield.*

Guidelines for Corn Losses

Remember that a few kernels on the ground do not mean the combine is doing a poor job. No combine is 100 percent efficient.

If total losses are acceptable, no adjustment of the machine or operating procedure is needed.

Check the losses occasionally to make sure that crop or field conditions haven't changed and increased losses. Changes in the moisture level, field conditions, how well the crop is standing and crop variety all affect the rate of loss.

Make only one adjustment at a time. Then check the losses. This will help you pinpoint the adjustment problem.

WEEDY FIELD CONDITIONS

Many field conditions affect the combine's efficiency. When special conditions are encountered, always check the grain loss to be sure it is held at a minimum.

Weedy fields increase losses and the moisture content of the harvested grain. Much of the heavy green material accumulates on the straw walkers, chaffer and sieves, and does not permit grain to be separated from the straw.

To reduce crop losses due to weeds, reduce the amount of green material taken into the machine by raising the cutter bar if the weeds are shorter than the crop. Go around especially weedy areas.

If many weeds must be taken into the combine to harvest the crop, some steps can be taken to reduce the crop loss:

1. First, slow down or take a narrower cut. This decreases the amount of material entering the machine and allows the walkers to clear the trash and separate the grain.

2. It may help to increase the cylinder clearance so

Fig. 84—Windrowing May Help Reduce Losses in Weedy Conditions

the green material is broken up as little as possible. However, this may result in overloading the straw walkers and can increase grain losses. If the cylinder spacing is increased, keep a careful check to see that the walkers are not overloaded.

3. On machines with fan air deflectors, change the direction of the fan blast so that a strong gust of air is present where the grain leaves the grain pan and passes on to the chaffer; this will aid in removing green material that is present with the grain.

4. Windrowing the crop may also be helpful (Fig. 84). The crop should be cut and windrowed about a week or ten days before a clean crop would be harvested. Using this procedure, the weeds will dry and less wet material will be taken into the machine. To assure proper drying, the windrow should be kept light and fluffy and resting on top of the crop stubble.

WEATHER CONDITIONS

Weather greatly affects the harvestability of the crop. If the crop moisture level is too dry at harvest time, harvesting in early morning or late afternoon when dew is present may help. The opposite, of course, is true for wet crop conditions.

Excessive moisture should be avoided. If snow or ice is present on the crop, the heat of the machine will melt the snow or ice and cause high grain moisture content. The moisture may also cause dust to turn to mud which can clog the straw walkers, sieves and chaffers. The same is true for rain or heavy dew.

CROP MOISTURE CONTENT

The moisture content of the crop not only affects the harvesting process, but also the price received for the crop and storage factors.

WHEAT should be harvested when the moisture level is below 14 percent. This permits safe storage of the wheat and selling without dockage due to excess moisture. Combining the wheat at a higher moisture

Fig. 85—Harvesting Soybeans When Dew Is On Them Helps Reduce Shatter Losses

Fig. 86—Lodged Crops Require Careful Combining

Fig. 87—Combining A Poor Windrow May Present Field Problems

level may damage the grain. The extra threshing action needed to separate a higher moisture crop, coupled with the softer grains, can crack the wheat.

SOYBEANS should be harvested with the moisture content below 14 percent, with the optimal level about 12 percent moisture. Most bean losses occur at the gathering unit due to shattering. If the beans can be harvested with some moisture on the pod, the shatter losses will be reduced. Early morning or late afternoon harvesting is best for dry beans, but care should be taken to avoid taking in too much moisture.

CORN should be harvested below 15 percent moisture for best storage results. Higher moisture content requires drying the grain and can increase the harvesting cost. Combining high moisture corn increases kernel damage at the cylinder because the crop is difficult to thresh and the kernels are softer. But drier corn has a higher preharvest loss. Both factors should be considered to determine the optimum time to harvest.

Variety also affects the optimum harvesting moisture. The best harvest moisture varies from operation to operation, but corn should not be harvested at moisture higher than 30 percent.

LODGED CROPS

When the crop to be harvested has lodged (is bent over or lying on ground), extra care must be taken in harvesting. Slower travel speeds and lower header heights help increase the percentage of the crop harvested. Special attachments are available to aid in harvesting down crops. Be careful to avoid damage from stones, sticks or other obstructions that may damage the machine when the header is operated at a lower height.

TROUBLESHOOTING OF FIELD PROBLEMS

The majority of combine operating problems can be traced to improper adjustment. The trouble shooting charts which follow will help you when problems develop by suggesting a probable cause and a recommended remedy. Always check the operator's manual of the combine you are operating for *specific* adjusting procedures and correct settings.

These suggested remedies should be applied with caution. Make an effort to understand the combine and know *why* you are making an adjustment. When you are trying to solve a problem, make certain that its source does not come from someplace *other* than the apparent cause. For example, a plugged cylinder may result from improper feeding at the feeder house, rather than an improperly adjusted cylinder.

To overcome harvesting problems, special attachments are available to meet various ground and crop conditions. But when you have difficulties, make every effort to correct the problem before purchasing a special attachment. If the combine is not adjusted properly, the special attachment will not eliminate the difficulty.

FIELD PROBLEMS

CUTTING PLATFORM PROBLEMS

Problem	Cause	Remedy
SHATTERING OF GRAIN AHEAD OF CUTTER BAR.	Reel speed not coordinated with ground speed, causing excessive agitation before crop is cut.	Change reel drive sprockets or adjust hydrostatic drive reel to coordinate reel speed with ground speed so reel will move crop smoothly and evenly.
		Reel should turn 25 percent faster than ground travel speed.
	Ground speed too fast for conditions of crops.	Slow down ground speed so reel will not bat crop, causing shattering.
CUT CROP BUILDING UP AND FALLING FROM FRONT OF CUTTER BAR OR LOSS OF GRAIN HEADS AT CUTTER BAR.	Reel not adjusted low enough for proper delivery of cut crop to auger.	Set reel low enough to sweep material from cutter bar.
	Reel set too far forward.	Set reel back closer to knife.
	Auger clearance too high from platform bottom.	Adjust outer ends of auger to 1/8- to 5/8- inch (3.2 to 15.6 mm) clearance of platform bottom and check finger clearance.
	Cutting platform too high; cutting stalks too short for proper delivery.	Lower cutting platform so stalks of cut material will be long enough for smooth, even feeding to auger.
	Reel speed too slow.	Increase speed of reel.
RAGGED AND UNEVEN CUTTING OF CROP.	Cutting mechanism not operating at recommended speed.	Check basic speed of combine (See Combine Operator's Manual.)
		Check platform and feeder house drives.
	Various parts of cutter bar, such as knife sections, guards, wearing plates, are worn, damaged, or broken.	Check and replace all defective parts on cutter bar to obtain even cutting of crop.
	Bent knife, causing binding of cutting parts.	Straighten the bent knife. Check guard alignment and align if necessary for a smooth cut.
	Knife clips not adjusted to permit knife to work freely.	Adjust knife clips so knife will work freely, but still keep knife sections from lifting off guards.
	Cutting edge of guards not close enough or parallel to knife sections.	Adjust guards.
	Looseness between knife back and guards.	Adjust wearing plates so knife back is snug to guard.
	Lips of guards out of adjustment or bent, causing poor shearing action.	Adjust lips of guards so they are parallel to shear edge of guards.
	Improper knife register.	Adjust knife register so knife sections pass an equal distance through adjacent guards at each end of pitman stroke.

(Continued on next page)

Problem	Cause	Remedy
CROP FALLING IN FRONT OF CUTTER BAR AFTER IT IS CUT.	Reel speed too slow.	Adjust reel speed to deliver crop to platform. Reel should turn 25 percent faster than forward travel of combine.
	Reel set too far forward.	Set reel back closer to knife.
	Cutting mechanism not at recommended speed.	Check basic speed of combine. (See Combine Operator's Manual.)
EXCESSIVE VIBRATION OF CUTTING PARTS.	Cutting mechanism not at recommended speed.	Check basic speed of combine. (See Combine Operator's Manual.)
		Check platform and feeder house drive. (See Combine Operator's Manual.)
	Excessive looseness of cutting parts and knife drive.	Remove all excessive play from cutter bar and knife drive to eliminate vibration. After removing excessive play, make certain cutter bar and knife drive are properly adjusted.
	Improper knife register.	Adjust knife register.
REEL WRAPPING IN TANGLED AND WEEDY CROPS.	Slat reel not efficiently delivering crop.	Install pickup reel.
	Incorrect location of reel.	Place reel well ahead and down. If wrapping occurs at ends of reel, install reel end shields.
	Reel speed too fast.	Reduce speed of reel to allow crops to fall into platform.
REEL CARRYING STRAW AROUND.	Reel speed too fast.	Reduce speed of reel so straw will not carry over top of reel. Reel should turn 25 percent faster than ground speed.
	Reel height too low.	Raise reel height to reduce amount of straw gathered by reel.
	Pickup tines pitched too much.	Reduce pitch of tines.
TOO MUCH MATERIAL ENTERING COMBINE.	Cutting too low in order to get all down and tangled crops.	Use lifting guards or pickup reel in down and tangled conditions.
	Ground speed too fast.	Reduce ground speed.
UNEVEN OR BUNCHED FEEDING OF CROP TO CYLINDER.	Auger clearance too high from platform bottom.	Adjust outer ends of auger to proper clearance of bottom.
	Build-up of grain on cutter bar.	Lower height of reel and set fore-and-aft position as close as possible to cutter bar and auger.
	Fingers in platform auger not adjusted to convey material properly to feeder conveyor chain.	Adjust finger height of auger so there is even feeding from the platform auger to feeder conveyor chain.
	Platform drive belt slipping.	Spring-loaded tightener must be free and tight against belt. (See Combine Operator's Manual.)
	Auger too far ahead of stripper.	Adjust auger back closer to stripper.
	Platform auger slip clutch set too loose.	Tighten auger slip clutch.
	Feeder drum lower stops set too high.	Adjust lower stops (See Combine Operator's Manual.)

ROW-CROP HEAD PROBLEMS

Problem	Cause	Remedy
PLUGGING ROW UNIT AT KNIFE.	Belt speed too high.	Reduce header speed.
LOSS OF CROP IN THE FIELD	Gatherer points set too high.	Lower gatherer points to pick up lodged or down crop.
	Ground speed too fast or too slow.	Adjust ground speed to meet field and ground conditions.
		Adjust speed of gathering belts.
	Not harvesting planter rows.	Harvest rows as they were planted. It will be easier to follow rows and reduce crop loss.
	Row units not centered on rows.	Adjust row unit row spacing to match row spacing of crop.
	Speed of gathering belts too fast or too slow.	Adjust speed of gathering belts.
	Skid shoes set to cut crop too high.	Adjust skid shoes for lower cut.
SOYBEAN PODS LEFT ON STUBBLE	Skid shoes set to cut crop too high.	Adjust skid shoes for lower cut.
	Excessive row unit float spring tension.	Adjust row unit float springs.
	Automatic header height control improperly adjusted.	Reset automatic header height control (see Operator's Manual).
	Gatherer points not properly positioned between rows.	Carefully operate row-crop head on rows; reduce ground speed if necessary.
POOR CUTTING ACTION (RAGGED AND UNEVEN CUTTING OF CROP)	Excessive gatherer belt speed.	Reduce gatherer belt speed.
	Incorrectly adjusted rotary knife.	Adjust rotary knife for proper cutting action.
	Bent, worn or broken rotary knife sections.	Straighten or replace rotary knife sections.
	Worn or broken stationary knife.	Reverse or replace stationary knife.
EXCESSIVE WEAR OR DAMAGE TO ROTARY KNIFE	Rocky field conditions.	Adjust skid shoes for higher cut.
		Be alert for obstructions during field operation.
	Rotary Knife is adjusted too close to stationary knife.	Adjust rotary knife for proper cutting action.
	Knife speed too high.	Reduce rpm to match ground speed.
EXCESSIVE WEAR OF STATIONARY KNIFE BLADES	Speed of gathering belts is too fast.	Reduce speed of gathering belts to match ground speed or increase ground speed.
	Excessive gap between rotary and stationary knife blades.	Be sure row unit frame and/or stationary knife are not bowed. Check mounting surface for stationary knife, to be certain it is free of weld spatter, paint, mud, and crop residue.
		Be sure all knife sections on each rotary knife contact the stationary knife evenly.
		Tap top end of rotary knife drive shaft with a brass or lead hammer to check that bearings are fully seated.
		Clean knives and adjust knife spacing (see Operator's Manual).

Problem	Cause	Remedy
WORN OR DAMAGED GATHERING BELTS OR GATHERING BELTS JUMPING OUT OF TIME	Dirt is forced into rotary knives by the skid shoes and gatherer belts.	Increase row unit float spring tension.
		Adjust skid shoes for a higher cut position.
		Adjust automatic header height control to activate sooner.
		Adjust gather points upward so they do not "funnel" dirt clods onto bean ridge.
	Inadequate gathering belt chain tension.	Adjust gathering belt chain tension.
	Gathering belts tear off grass, weeds, or crop before they can be cut.	Reduce speed of gathering belts to match ground speed or increase ground speed.
	Rocky field conditions.	Install rock deflectors.
	Poor cutting by rotary knives.	Adjust rotary knives for proper cutting action.
		Reduce speed of gathering belts.
	Trough under gathering belts fills with dirt and cut stalks.	Increase row unit float spring tension.
		Adjust skid shoes for higher cut.
		Adjust automatic header height control to activate sooner.
		Adjust gather points so dirt is not funneled into bean ridge.
	Inadequate gatherer belt chain lubrication.	Lubricate gatherer belt chains.
DIRT IN GRAIN TANK OR DIRT PLUGS TROUGH UNDER GATHERING BELTS	Excessive pressure on skid shoes causes them to push dirt.	Increase row unit float spring tension.
		Operate row-crop head as high as possible so skid shoes just touch the ground.
		Adjust automatic header height control to activate sooner.
		Adjust gather points so dirt clods are not funneled into bean ridge.
	Gathering belts pull grass, weeds, or crop out of the ground so dirt is brought in with the roots.	Reduce speed of gathering belts to match ground speed or increase ground speed.
		Adjust rotary knives for proper cutting.
	Rows are ridged high enough to allow gathering belts or rotary knives to contact dirt.	Adjust skid shoes for higher cut position.
	Dirt not being removed by combine.	Use perforated elevator doors, auger connections, and feeder house bottom insert when combining soybeans, edible beans and similar crops to help remove dirt and weed seed.
SKID SHOES DIG INTO GROUND	Insufficient float spring tension.	Increase row unit float spring tension.
	Automatic header height control improperly adjusted.	Reset automatic header height control.
SKID SHOES WILL NOT FLOAT PROPERLY	Excessive skid shoe pressure on ground.	Increase row unit float spring tension.
		Adjust automatic header height control.

Problem	Cause	Remedy
ROW UNITS DO NOT MOVE FREELY THROUGH FLOAT RANGE	Insufficient clearance or binding between gatherer sheets.	Adjust gatherer sheet clearance between row units.
	Row units do not move freely on pivots.	Lubricate row unit pivots at grease fittings provided.
	Trash collects at row unit pivots.	Clean around row unit pivots.
ROW-CROP HEAD IS TOO SLOW TO RETURN TO OPERATING LEVEL AFTER ENCOUNTERING IRREGULAR GROUND	Excessive float spring tension.	Adjust row unit float spring tension.
	Insufficient clearance or binding between gatherer sheets.	Adjust gatherer sheet clearance between row units.
	Automatic header height control improperly adjusted.	Adjust drop rate control.
SKID SHOES WILL NOT FLOAT PROPERLY	Excessive skid shoe pressure.	Increase float spring tension.
		Adjust header height switches to operate in lower part of range.
SKID SHOES DIG INTO GROUND	Insufficient float spring tension.	Increase float spring tension.
	An individual row unit does not signal a change in ground contour.	Adjust actuators on individual row units.
ROW UNITS DO NOT MOVE FREELY THROUGH FLOAT RANGE	Insufficient clearance or binding between gatherer sheets.	Adjust gatherer sheet clearance between row units.
	Row units do not move freely on pivots.	Lubricate row unit pivots at grease fittings provided.
	Trash collects at row unit pivots.	Clean around row unit pivots.
HEADER IS TOO SLOW TO RETURN TO OPERATING LEVEL AFTER ENCOUNTERING IRREGULAR GROUND	Excessive float spring tension.	Adjust float spring tension.
	Insufficient clearance or binding between gatherer sheets.	Adjust gatherer sheet clearance between row units.
	Drop rate valve slows drop rate more than desired.	Readjust drop rate valve.
	Deck plates not adjusted properly.	Adjust deck plates.
	Corn head set too high.	Lower corn head.
CORN HEAD PROBLEMS		
LOSS OF EARS FROM THE HEAD.	Gatherer points set too high.	Adjust gatherer points so they just touch the ground. When picking low hanging ears, raise front tip of gatherer points just enough to run the corn head with skids close to the ground.
	Ground speed too fast or too slow.	Operate at a speed to meet field and ground conditions. Excessive ground speed can cause ears to fall off the stalks ahead of the gatherer chains. Too-slow ground speed can cause the ears to slide forward out of the unit. Operate at a speed where the gatherer chains merely help guide the stalks into the rolls.
	Not picking planter rows.	Pick rows as they were planted. It will be easier to follow the rows and eliminate loss of ears.

Problem	Cause	Remedy
EAR SHELLING AT STALK ROLLS.	Row units not centered on rows.	Adjust corn head row spacing to equal row spacing of corn.
	Ears sliding out over gatherer chains.	Use ear savers and center shield extensions.
	Gatherer chain speed too fast or too slow.	Obtain the correct gatherer chain speed by changing the feeder house powershaft drive sprocket.
	Deck plates not adjusted properly.	Adjust deck plates.
	Corn head set too high.	Lower corn head.
EARS NOT SHELLED COMPLETELY.	Moisture content of corn too high.	Wait for moisture content of corn to drop. Corn kernels tend to cling to the cobs when the moisture content is above 30 percent. Best shelling is obtained and crackage is at a minimum when the moisture content is under 27 percent.
	Cylinder or rotor speed too slow.	Increase cylinder or rotor speed by 5 percent and *check combine primary countershaft speed* or combine beater speed. Also, check separator drive belt tension to make sure it is not slipping. (See Combine Operator's Manual.)
	Rasp bars bent.	Straighten or replace rasp bars.
	Concave bent.	Replace if necessary.
	Concave not level. (Not parallel with cylinder or rotor.)	Adjust concave spacing equally on both ends of cylinder or rotor.
	Too wide a space between cylinder or rotor and concave.	Close cylinder-to-concave (or rotor-to-concave) spacing to increase shelling action.
	Ears going between rasp bars of cylinder without being shelled.	Install cylinder filler plates.
	Cobs being split without the corn being shelled from them. (Corn is attached to half or smaller section of cob.)	Open concave spacing just enough to get proper shelling action.
	Ground speed too fast.	Reduce ground speed.
EXCESSIVE DAMAGE TO SHELLED CORN-CRACKED CORN.	Concave too close to cylinder or rotor bars.	Increase concave spacing.
	Cylinder or rotor speed too fast.	Reduce cylinder or rotor speed. Also check combine beater speed or combine primary countershaft speed as shown in Combine Operator's Manual.
	Moisture content of corn too high.	Wait until moisture content of corn drops. Corn above 30 percent moisture has a tendency to crack and is easily crushed. It is best to wait until moisture content is under 27 percent.
	Concave not level.	Level concave.
	Damaged rasp bars or concave.	Replace as necessary.

(Continued on next page)

Problem	Cause	Remedy
	Dented auger housings.	Straighten or replace as necessary.
	Excessive tailings.	Reduce ground speed.
		Open or clean the sieve and increase the fan speed.
SHELLED CORN COMING OUT REAR OF COMBINE.	Corn carrying over straw walkers.	Extend pans at rear of straw walkers, if straw walkers are so equipped.
		Clean out straw walkers if they are plugged with cobs.
	Corn carrying over chaffer.	Reduce ground speed.
		Adjust the chaffer if too far closed or too far open and plugged with pieces of cob. Refer to Combine Operator's Manual.
		Clean sieve completely if sieve is closed or clogged with cobs.
		Increase the speed of cleaning fan if volume of air does not appear to be adequate. Check belt tension.
		If combine is equipped with sieve tilt adjustment, lower front of sieve to allow more air to the chaffer.
		Too much material in combine. Check corn head for excessive stalk breakage which could be due to rolls not being properly timed or deck plates closed too much.
EARS SLIDING OUT THROUGH THE THROAT.	Ear savers not properly adjusted.	Adjust ear savers.
	Ear savers worn out.	Replace ear savers.
PULLING UP CORNSTALKS.	Deck plates set too close together.	Spread deck plates, a little at a time, until stalks feed through rolls more freely.
	Traveling too fast for gatherer chain speed.	Slow down to meet crop conditions or increase row unit drive speed.
	Gatherer chain flights digging into cornstalks roots.	Lower gatherer points.
	Corn extremely dry or down.	Remove center shield extensions and ear savers.
	Worn stalk rolls.	Replace stalk rolls.
PLUGGING.	Stalks breaking in stalk rolls or deck plates.	Adjust the opening of deck plates. Check the stalk roll timing so stalk roll flutes do not break stalks. Also make sure deck plates are set equidistant.
	Trash winds around stalk rolls.	Set trash knives closer to stalk rolls.
	Loose gatherer chains.	Check gatherer chain mechanism.
	Not picking planter rows.	Pick rows as they were planted. It will be easier to follow the rows, reduce plugging, and eliminate loss of ears.

(Continued on next page)

Problem	Cause	Remedy
	Material catching on sheet metal.	Check for broken or bent sheet metal that may prevent flow of material.
	Ground speed too fast, causing too much material to go into corn head too fast.	Slow down. Operate at a speed to meet the yield and ground conditions. Faster speeds will cause plugging.
	Material not flowing through cross auger.	Check for obstructions in cross auger housing and for roughness on auger.
	Corn stalks plugging in gatherer throat opening.	Remove ear savers and center shield extensions.
COBS AND FOREIGN MATERIAL IN GRAIN TANK.	Moisture content of corn too high.	Check moisture content of corn before harvesting. Do not harvest if over 30 percent moisture.
	Insufficient air blast from cleaning fan.	Increase fan speed to obtain sufficient air blast. Lower front of sieve. Keep sieve clear of pieces of cob and obstructions.
WINDROW PICKUP PLATFORM PROBLEMS		
WINDROW DIFFICULT TO PICK UP OR STRIPS OF CROP NOT BEING PICKED UP.	Short crop down in the stubble embedded in furrows.	Raise gauge wheels to allow fingers to run closer to ground.
	Gauge wheels not always contacting ground.	Lower platform to allow pickup to operate at less pitch.
	Pickup speed too slow.	Increase pickup speed.
PICKING UP DIRT.	Improper gauge wheel setting.	Lower gauge wheels to prevent pickup fingers from digging in ground.
	Pitch of pickup too flat.	Raise platform so area between center and drive rollers is parallel with ground.
	Excessive pickup speed.	Reduce speed of pickup.
PICKING UP STONES.	Improper gauge wheel setting.	Lower gauge wheels to increase finger clearance to ground.
	Excessive pickup speed.	Reduce speed of pickup.
	Pitch of pickup too flat.	Raise platform to increase pitch.
PICKUP FEEDS TOO HIGH ON AUGER.	Excessive pitch on pickup.	Lower platform so area between center and drive rollers is parallel with ground.
	Hold-down fingers not adjusted properly.	Lower hold-down fingers to provide more compression of crop at rear of pickup.
	Excessive pickup speed.	Reduce speed of pickup.
WINDROW ROLLS AHEAD OF PICKUP.	Speed of pickup too slow.	Increase speed of pickup.
MATERIAL BUILDS UP AT PICKUP STRIPPER OR BUILDS UP BETWEEN PICKUP STRIPPER AND AUGER.	Speed of pickup too slow.	Increase speed of pickup.
	Platform auger not adjusted properly.	Adjust platform auger down. Check auger finger adjustment. (See combine operator's manual.)

FEEDER HOUSE PROBLEMS		
Problem	**Cause**	**Remedy**
UNEVEN OR BUNCHED FEEDING OF CROP TO CYLINDER.	Auger clearance too high from platform bottom.	Adjust outer ends of auger to proper clearance of bottom. See Operator's Manual.
	Build-up of grain on cutter bar.	Lower height of reel and set fore-and-aft position as close as possible to cutter bar and auger. See Operator's Manual.
	Front of feeder conveyor chain adjusted too high.	Adjust feeder drum so conveyer clears bottom by 1/8 inch.
	Feeder conveyor chain too tight and holds drum up.	Adjust conveyor chain to proper tension.
	Fingers in platform auger not adjusted to convey material properly to feeder conveyor chain.	Adjust finger height of auger so there is even feeding from the platform auger to feeder conveyor chain. See Operator's Manual.
	Platform drive belt slipping.	Spring-loaded idler must be free and tight against belt.
	Auger too far ahead of stripper.	Adjust auger back to stripper. See Operator's Manual.
	Platform auger slip clutch set too loose.	Tighten auger slip clutch. See Operator's Manual.
	Feeder drum swing arms binding and not letting drum down. Spring mechanism on arms not functioning.	Check inside of feeder house and remove any material buildup from around swing arms and spring.
	Feeder conveyor slats bowed.	Straighten or replace bent slats.
	Feeder running too fast or too slow.	Adjust feeder speed.
CYLINDER OR ROTOR SPEED PROBLEMS		
EXCESSIVE DROP IN CYLINDER OR ROTOR SPEED.	Basic separator speed not correct.	Check primary countershaft speed with engine at full throttle, no load.
	Improper adjustment of primary countershaft drive belts.	Tighten primary countershaft drive belts.
	Improper adjustment of cylinder or rotor drive belts.	Tighten drive belts to recommended tension.
	Engine governor not operating properly.	Adjust governor. See Operator's Manual.
CYLINDER OR ROTOR SPEED WILL NOT ADJUST.	Threads on adjusting screws fouled or dirty.	Clean and adjust speed control mechanism.
	Separator not running.	Start separator.
	Variable sheaves out of time.	Time sheaves. See Operator's Manual. *(Continued on next page)*

Problem	Cause	Remedy
CYLINDER OR ROTOR TURNING BUT NO READING ON THE TACHOMETER.	Broken tachometer shaft.	Replace flexible shaft.
	Broken or loose drive belt on tachometer.	Replace drive belt.
CYLINDER OR ROTOR NOT OBTAINING FULL SPEED RANGE.	Improper primary countershaft speed.	Adjust primary countershaft speed with engine at full throttle, no load.
	Variable sheaves out of time.	Time sheaves.
THRESHING PROBLEMS		
SLUGGING OR OVERLOADING OF CYLINDER OR ROTOR	Basic separator speed not correct.	Check primary countershaft speed with engine at full throttle, no load.
	Engine not up to correct speed.	Adjust governor to correct gasoline engine speed.
		Adjust diesel injection pump governor.
	Separator belt slipping.	Adjust separator belt to proper tension.
	Variable speed cylinder or rotor drive belts slipping.	Adjust variable speed cylinder or rotor drive belts to proper tension.
	Concave spacing too close.	Open concave spacing to maintain proper threshing.
	Cylinder or rotor speed too slow.	Increase cylinder or rotor speed.
	Too much material entering cylinder or rotor.	Reduce ground travel speed.
BACKFEEDING OF CYLINDER OR ROTOR	Engine not up to correct speed.	Adjust governor to correct gasoline engine speed.
		Adjust diesel engine speed.
	Basic separator speed not correct.	Check primary countershaft speed with engine at full throttle, no load.
	Straw walker slip clutch slipping.	Determine cause of slipping and correct.
	Separator curtain too close to rotary deflector.	Remove front curtain.
	Rotary deflector not turning.	Check rotary deflector drive belt.
GRAIN NOT THRESHED FROM HEADS	Crop not in condition to thresh.	Test moisture content of crop before combining.
	Cylinder or rotor speed too slow.	Increase cylinder or rotor speed enough to do a good job threshing, do not increase speed to the point where grain cracks.
	Concave spacing too wide.	Close concave spacing to increase threshing action.
	Unthreshed heads passing through concave grate.	Install concave cover plates to keep unthreshed heads from passing through grate.
	Uneven feeding to cylinder or rotor.	Check feeder conveyor chain tension and float in feeder house.
	Not enough material entering combine for proper threshing.	Increase ground travel speed for more intake of material.

Problem	Cause	Remedy
EXCESSIVE CRACKED GRAIN IN GRAIN TANK.	Cylinder or rotor speed too fast for crop.	Decrease cylinder or rotor speed just enough to stop grain cracking, but still do a good threshing job and/or open concave slightly.
	Concave spacing too close.	Open concave spacing just enough to eliminate cracking. Decrease cylinder or rotor speed.
	Uneven feeding or slugs entering cylinder or rotor.	Check feeder conveyor chain tension and float in feeder house.
	Excessive clean grain in tailings, causing grain to crack when rethreshed.	Open sieve slightly to reduce tailings. Lower front of sieve.
	Not enough straw entering combine.	Increase ground speed to increase amount of material being taken into combine.
	Dented auger housings or bent auger shafts cracking grain between flights and housings.	Remove dents from auger housings and/or straighten bent auger shafts to eliminate cracking.

SEPARATING PROBLEMS

Problem	Cause	Remedy
MATERIAL LODGING ON STRAW WALKER, AND NOT BEING EVENLY DISCHARGED FROM REAR OF COMBINE.	Over-all separator speed too slow for proper action of straw walkers.	Check primary countershaft speed. Adjust gasoline engine governor. Adjust diesel engine speed. Check separator drive if necessary.
	Walker slip clutch slipping.	Determine cause of slipping and correct.
	Material catching on straw walker curtain and building up on front of walkers.	Remove front curtain.
	Material not being delivered evenly to walkers.	See section on improper cylinder action.
GRAIN LOSS OVER STRAW WALKERS.	Straw walkers running at incorrect speed.	Check primary countershaft speed with engine at full throttle, no load.
	Damaged straw walker curtains.	Install new curtains.
	Straw walker overloaded due to incomplete threshing or late threshing at the concave.	Reduce concave spacing and/or increase cylinder or rotor speed to increase threshing action.
	Straw walker openings plugged so threshed grain cannot drop through to conveyor augers.	Clean out openings in straw walkers.
	Straw walkers overloaded so grain cannot get through material to conveyor augers.	Reduce ground travel speed to reduce amount of material entering combine. Check to be sure curtains are in place. Raise cutting platform to cut less material.
	Material overthreshed and pulverized. Cylinder or rotor speed too fast.	Reduce cylinder or rotor speed.
	Mat of straw not being broken up.	Remove one riser from same side of each straw walker.

CLEANING PROBLEMS

Problem	Cause	Remedy
FOREIGN MATERIAL IN CLEAN GRAIN.	Basic speed of combine incorrect.	Check primary countershaft speed with engine at full throttle, no load.
	Insufficient air blast from cleaning fan.	Increase fan speed.
	Sieve lips open too far, allowing foreign material to fall through with clean grain.	Consider more air blast; then close sieve lips so foreign material will be carried back to the tailings auger.
	Cleaning shoe overloaded with fine, chopped straw.	Open concave spacing to reduce amount of broken straw. Increase air blast. Reduce cylinder or rotor speed. Position sieve in upper position.
	Air deflectors or wind boards improperly adjusted.	Adjust air deflectors or wind boards.
LOSS OF GRAIN OVER CLEANING SHOE.	Basic speed of combine incorrect.	Check basic speed of separator at primary countershaft with engine at full throttle, no load.
	Chaffer overloaded, causing carryover of grain.	Increase air blast. Open lips of chaffer.
	Grain blowing over shoe.	Reduce fan speed.
	Too much broken straw on chaffer for proper cleaning.	Open concave spacing and/or reduce cylinder or rotor speed. Reduce ground travel speed.
EXCESSIVE CLEAN GRAIN IN TAILINGS BEING RETURNED TO CYLINDER.	Incorrect setting of adjustable sieve for condition of crop.	Increase sieve opening to permit clean grain to fall through sieve before it is carried back to tailings auger.
	Incorrect setting of fan blast for condition of crop.	Reduce fan speed to permit clean grain to fall through sieve before it is carried back to tailings auger.
		If increasing sieve openings and/or reducing fan speed results in foreign material in the clean grain, then lower the front of the sieve to the bottom hole and readjust the sieve openings and the fan speed.
EXCESSIVE TRASH IN TAILINGS.	Insufficient air blast.	Increase fan speed.
	Chaffer lips too far open.	Close chaffer and/or extension lips slightly.
	Overthreshing.	Open concave spacing and/or reduce cylinder or rotor speed.

CHAPTER QUIZ

1. What are five clear signs of poor combine harvesting?

2. What are four things a combine operator should do prior to the harvest season when planning and preparing for harvest?

3. (Fill in blank.) Turning, refueling, breakdowns, etc. will account for about _____ percent of the harvesting time during combine operation.

4. Preliminary settings on a combine are very important. What four areas could be set during the off season as a starter for proper combine adjustment?

5. In small grains, what four losses would you check to see if the combine was set correctly?

6. Identify the results of improper threshing. Place an "O" beside the items below which indicate *overthreshing* and a "U" beside items which indicate *underthreshing*.

    ___A. Grain losses over shoe

    ___B. Broken and chewed straw

    ___C. Overloaded straw walkers

    ___D. Cracked kernels

    ___E. Unthreshed heads

    ___F. Grain losses over walkers

7. What are six sources of grain losses?

8. (Fill in blank.) A _____ spreading device should not be used when checking grain losses.

9. True or false? "Acceptable losses in small grains range from five to eight percent."

10. (Fill in blank.) The majority of combine operating problems can be traced to improper _____.

6
Maintenance and Safety

Fig. 1—The Combine Requires Regular Maintenance and Service Adjustments

MAINTENANCE

Proper maintenance and service adjustments are necessary to assure efficient, safe operation of the combine. Costly repairs, premature wear, loss of field time and accidents can be reduced if the combine is properly maintained and adjusted. The fundamentals of preventive maintenance are covered in manuals listed in Suggested Readings section — page 210.

The operator's manual for the machine should be used in reference to *specific* maintenance intervals, location of service points and instructions for the performance of maintenance and service adjustments. Because of differences in makes and models of combines, this chapter deals only with general service and maintenance requirements. Always study the operator's manual carefully to determine what maintenance is needed.

GENERAL MAINTENANCE

There are several practices that a good machine operator always follows. He knows that by following these rules, his job of operating and maintaining the combine will be much easier and safer.

1. *Always keep the machine clean.* Before starting the combine, clean all field trash, mud and excess grease and oil from the machine. Not only is this

Fig. 2—Keeping Maintenance Records

Fig. 3—Checking Engine Oil Level

a good safety practice, it also helps the combine to run more efficiently, prevents moisture accumulation and rust on metal parts and cuts down on time lost in the field for repairs.

2. *Make sure that nuts, cap screws, shields and sheet metal parts are tight.* A loose shield can vibrate, produce irritating noise, and cause a machine failure if it falls in the way of moving parts. Loose attaching hardware can cause breakdowns that take time the machine should be using for work.

3. *Inspect the combine before starting every day.* A brief look at all areas of the combine can help you spot potential machine failures and safety hazards.

4. *Keep maintenance records (Fig. 2).* A simple chart showing when lubrication and service adjustments were made can help you make sure that all needed maintenance has been performed.

5. *Don't abuse the machine.* Proper lubrication and adjustment is of little help if the operator abuses the machine. A good combine operator follows his operator's manual and makes sure he doesn't overload the machine, operate it at speeds too fast for field conditions, or operate it under conditions that could cause damage to the machine.

ENGINE AND POWER TRAIN MAINTENANCE

Before starting the combine each day, the following checks should be made:

1. Check the oil level of the engine crankcase (A, Fig. 3).

Fig. 4—Checking Engine Fan Belt Alignment and Tension

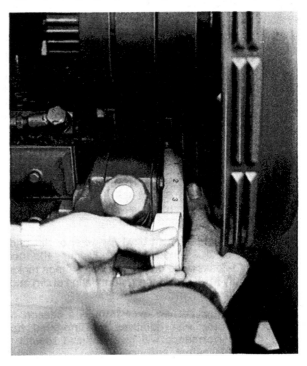

183

If the level is low, be sure to add the proper amount and type of engine oil (B, Fig. 3). Refer to the operator's manual instructions as to how often the oil and filter should be changed.

2. Clean off any accumulation of grease, excess oil or dirt. This will help the engine to run cooler and more efficiently.

3. Check the level of coolant in the radiator daily. If the level is low, add the proper type of coolant. In cold weather, make sure the engine has enough anti-freeze to prevent freeze up. Remove any trash that has accumulated on the radiator or air intakes. Use compressed air to blow out dust and leaves so enough air can pass through the radiator to cool the engine efficiently.

4. Make sure engine fan belts are in good condition and tightened properly (Fig. 4).

5. Check the electrolyte level in the battery (Fig. 5). Running a battery with electrolyte too low will shorten its life and reduce its power output. Inspect battery terminals and all other cable connections. Make sure they are tight and free of corrosion.

6. Periodically inspect the ignition system on spark-

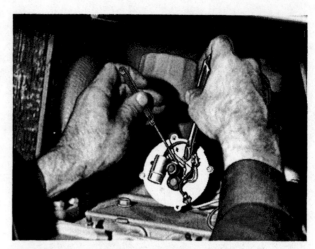

Fig. 6—Checking Distributor Breaker Point Gap (Spark-Ignition Engines)

8. Check the hydraluic fluid levels and add fluid if needed. Be careful to wipe all dirt away from the inspection point before checking the level to prevent dirt from entering the system.

Use a piece of cardboard or wood to check hydraulic lines for leaks. High pressure leaks are sometimes invisible, but still can penetrate the skin. Be careful to avoid injury.

If the combine has a separate power steering reservoir, check the fluid level.

9. Check the brakes for even brake pedal pressure as specified in the operator's manual.

10. Make sure the transmission and drive units have the proper amount of lubricant. Clean the inspection areas before checking lubricant levels.

11. Check variable-sheave and drive belts for proper alignment. Examine hydrostatic speed range lever to see that correct linkage and adjustment is maintained.

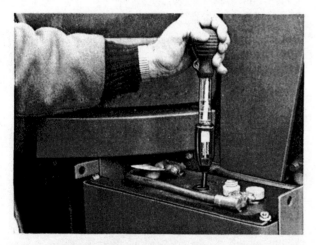

Fig. 5—Checking Battery Electrolyte Level

Fig. 7—Inspect the Fuel Sediment Bowl

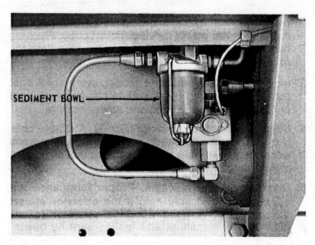

ignition engines. Clean and adjust the distributor breaker points (Fig. 6) and spark plugs as needed. Watch for loose spark plug wires and make sure the insulation is in good condition.

7. Check the fuel sediment bowl for water or dirt and drain and clean as needed (Fig. 7). Examine the fuel system for leaks. Also drain and clean storage tanks periodically to eliminate the accumulation of dirt and water condensation.

Never use fuel carried over from one season to another. Fuels with additives for improved hot weather performance will give reduced results if used in cold weather, and vice versa.

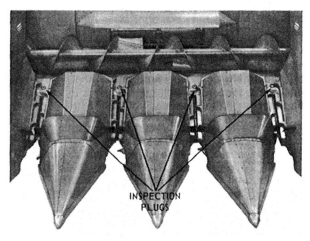

Fig. 8—Check Lubricant Level in Corn Head Gear Boxes

12. Check the clutch pedal for proper free travel as specified in the operator's manual.

13. Periodically inspect wheel bearings. Clean and repack them as specified in the operator's manual.

14. *Remember: Lubricate all points specified in the operator's manual.*

HEADER MAINTENANCE

Before operating the header, perform the following procedures.

LUBRICATION

Lubricate all header chains at frequent intervals. Operate chains for several minutes so they are warm when oil is applied. (Proper chain and belt maintenance is discussed later in this chapter.)

Fig. 9—Check Cutterbar Wear Plates for Correct Alignment

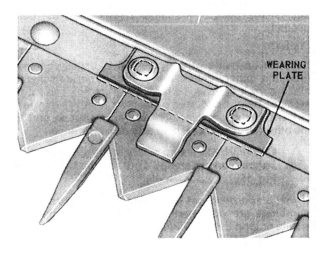

- *Lubricate the reel shaft daily with a multi-purpose grease.*
- *Lubricate the cutterbar drive daily.*
- *Check gear cases on corn head for proper lubricant level (Fig. 8).*
- *Lubricate the cutterbar knife as needed.*

INSPECTION

1. Check the cutterbar frequently for:
- *Broken sections*
- *Proper alignment of knife bar*
- *Proper alignment of guards*
- *Proper register*

2. Also make sure cutterbar wear plates and knife clips are in good alignment (Fig. 9).

3. Check the knife blade for a sharp cutting edge.

4. Make sure the knife stroke is the proper length.

5. Check to see that all belts and chains are tightened properly and that the flights on the feeder conveyor are not loose, bent or broken (Fig. 10).

6. Inspect all sealed bearings for excessive play and replace if necessary.

7. Inspect the header to see it is properly aligned and level.

Fig. 10—Check for Loose or Broken Flights on Feeder Conveyor

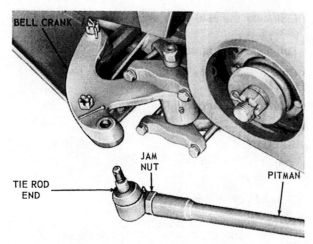

Fig. 11—Tie Rod End on Pitman is used for Knife Register Adjustment

PLATFORM ADJUSTMENT

If the platform requires adjustment, perform the following steps.

Cutter Bar

To remove the knife for sharpening or repair, disconnect the cutterbar drive and pull the knife out of the cutterbar. Be very careful if it is necessary to drive the knife out of the cutterbar. Always use a wooden block or rubber mallet to strike the knife. Use care to avoid bending the knife.

To avoid incomplete cutting of the crop and frequent clogs, it may be necessary to adjust the cutterbar register. When the register is correct, the knife sections pass an equal distance through adjacent guards at each end of the pitman stroke. Adjusting the register is usually done by lengthening or shortening the pitman stroke.

Adjust at the tie rod end, usually located between the pitman and the bell crank (Fig. 11). Make sure that adjustment nuts are tight after adjusting the tie rod.

Proper guard alignment is necessary for proper shearing of the crop. Guards that are bent or out of adjustment can also damage the knife. To adjust the guard, loosen the bolt connecting it to the cutterbar, then align the guard and tighten the bolt (Fig. 12).

Knife clips must keep sections from lifting off the guards and also permit the knife to slide without binding. Always set the knife clips *after* the guards are aligned. To set the clips, tap them up or down with a hammer until the knife will slide under them without binding (Fig. 13). Never bend the knife clip down when a knife section is directly under it.

Wear plates are located along the entire length of the knife

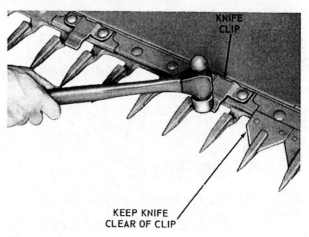

Fig. 13—Adjusting Knife Clips

Fig. 12—Adjusting Knife Guards

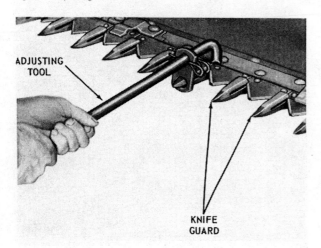

back and are adjustable to compensate for wear on the knife back. The wear plates must line up with each other to give the knife back a straight bearing along its entire length.

To set the wear plates, loosen the bolts on the knife clip, hold the knife forward and adjust the plates finger tight against the back of the knife. Then tighten the bolt securely.

Platform

Because one side of the platform is usually heavier than the other, check the platform periodically to make sure it is level. To level the platform, raise it to medium height and take a position about fifteen feet in front of the combine. Then sight from one end to the other, using the combine axle as a reference.

 CAUTION: Make sure safety stops are in position before working around or under the header (see Fig. 15).

If the platform is not level, adjust as follows—Most combines are equipped with either slotted adjusting points or long threaded bolts with adjusting nuts.

Fig. 14—Adjusting Bolts for Leveling Platform

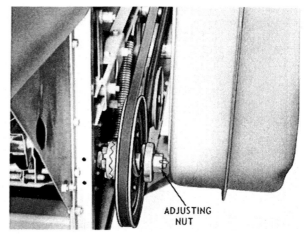

Fig. 16—Slip Clutch

If the combine is equipped with a slotted adjustment, loosen the bolts (Fig. 14) and level the platform. Then re-tighten the bolts.

If the combine is equipped with a bolt adjustment, loosen the jam nuts and then tighten or loosen the adjusting nuts as needed. After adjustment, make sure jam nuts are tight against the adjusting nuts.

It may also be necessary to align the cutterbar with the platform. The support for the cutter bar under the combine may move, causing the cutterbar to be lower at one end than the other, in relation to the platform.

To align the cutterbar, lower the platform onto a wooden block. Then loosen the bolts on the support arms, slide the cutter bar up or down to align it with the platform and tighten the bolts.

Always level the platform before aligning the cutterbar.

Auger Slip Clutch

The platform auger slip clutch is designed to prevent damage to the platform auger should it become clogged or if an obstruction enters the platform. The slip clutch springs should be set with just enough tension to work without slipping. Never increase the spring tension so much that the clutch will not slip.

To adjust the clutch, remove the cotter pin holding the adjusting nut and tighten or loosen the nut as needed (Fig. 16). Set the clutch loosely at first and then tighten it gradually until the proper working tension is determined. Then replace the cotter pin.

ROW-CROP HEAD ADJUSTMENT

Proper adjustment of the row-crop head reduces crop losses, saves operating time, and can help cut repair bills.

Fig. 15—Always Make Sure Safety Stop Is In Position Before Working Under Header

Fig. 17—Keep Chains Adequately Tensioned

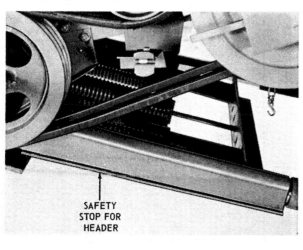

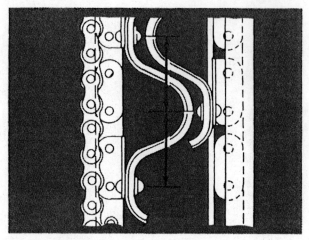

Fig. 18—Keep Gathering Belts Properly Timed

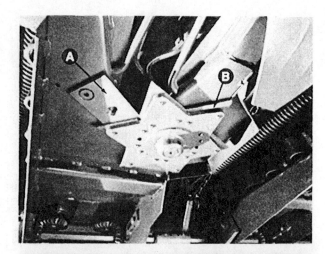

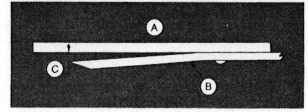

Fig. 19—Stationary And Rotary Knives Must Be Sharp

Drive Chain Tension

Keep all drive chains tight enough to keep chain links from climbing or jumping over sprocket teeth (Fig. 17). But, avoid overtightening which increases wear on chains, sprockets, bearings, and shafts. See the operator's manual for proper tension on each chain.

Timing The Gathering Belts

Gathering belts must be correctly timed to properly grip crop stalks and feed them to the rotary knives. Belt timing can be disturbed if one of the belt drive slip clutches is activated. Always recheck gathering-belt timing whenever an obstruction has caused the slip clutches to slip. To retime the belts, loosen belt drive chain tension and reposition the belt drive chain so that the belt loops are properly mated (Fig. 18).

Cutting Knives

Knives must be sharp and clean for smooth, efficient cutting, especially in weedy conditions, and each section of the rotary knife must contact the stationary knife evenly. When the stationary knife (A, Fig. 19) becomes badly worn, reverse or replace it. As clearance (C) between stationary knife (A) and rotary knife (B, Fig. 19) becomes greater, the risk of uneven cutting and plugging increases. To reduce clearance between knives, remove the stationary knife and install special shims before replacing the knife. Replace dull knife sections on the rotary knife by removing the entire knife, taking off old sections, and riveting on new ones.

Auger Settings

Set the clearance under the cross auger at approximately ¼ to ½ inch (6 to 13 mm) for soybeans and crops with smaller seed pods and stalks (Fig. 20), but increase clearance to as much as 1 inch (25 mm) for sunflowers and sorghum. Maintain proper clearance between the auger and auger stripper across the back of the head to prevent the crop from carrying over or wrapping around the auger.

Slip Clutches

Slip clutches protect equipment from damage in case of plugging or sudden overload. However, if slip clutches slip too easily, time is wasted and plugging is increased. Therefore, tighten all slip clutches enough to prevent unnecessary slippage without overtightening (Fig. 21). If clutch slippage continues to be a problem, determine and correct the cause rather than overtighten the clutch which would then eliminate protection from damage.

Fig. 20—Set Auger And Stripper Clearance For Crop Being Harvested

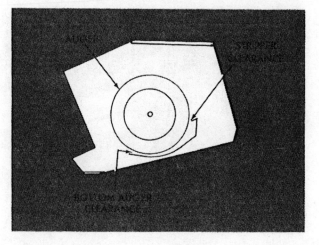

Fig. 21—Keep Slip Clutches Properly Adjusted

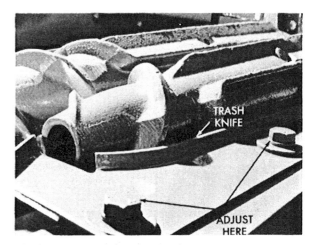

Fig. 23—Adjusting Bolts for Trash Knives

CORN HEAD ADJUSTMENT

The following paragraphs describe adjustments that may be required on a typical corn head.

Gatherer Chains

Proper tension on the gatherer chains should be maintained at all times. Most heads are equipped with adjustable tighteners. To adjust the tension, turn the adjusting nut (Fig. 22). To remove the chain, loosen the adjusting nut until the chain can be removed. See the operator's manual for specific procedures.

Trash Knives

Most trash knives have slotted holes that permit adjustment (Fig. 23). Loosen the bolts holding the knives in place and slide the knives until the proper cutting gap is reached. Then tighten the bolts.

FEEDER CONVEYOR ADJUSTMENTS

Chain

The feeder conveyor should be kept in adjustment to assure proper handling of the crop. Before adjusting the conveyor chain, remove the cover and clean away all trash. Most combines are equipped with eyebolt adjustments, located on the outside of the conveyor housing (Fig. 24). Tighten or loosen these bolts to obtain the correct tension. Make sure that both sides are adjusted equally.

THRESHING UNIT MAINTENANCE

The following paragraphs describe maintenance of a typical threshing unit.

LUBRICATION

Because makes and models vary greatly in lubrication

Fig. 22—Adjusting Gatherer Chain Tension

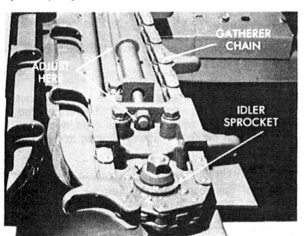

Fig. 24—Eyebolt Adjustment for Feeder Conveyor Chain Tension

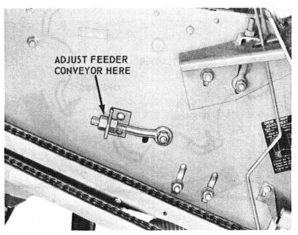

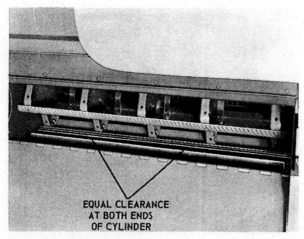

Fig. 25—Make Sure Concave Clearance Is Equal at Both Ends of Threshing Cylinder Or Rotor

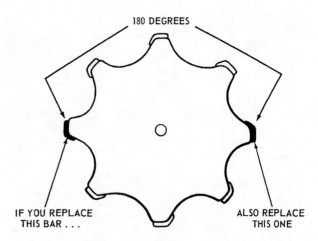

Fig. 27—If One Rasp Bar Needs Replacement, You Must Also Replace The Bar 180 Degrees Opposite To Maintain Cylinder Balance

requirements, always check the operator's manual to determine lubrication points and intervals and recommended lubricants.

INSPECTION

1. Check all belts and chains for proper tension and for excessive wear. Refer to belt and chain maintenance procedures later in this chapter.
2. Examine the condition of the threshing cylinder or rotor bars periodically. Make sure cap screws and nuts are tight.
3. Check the alignment and clearance of spike-tooth cylinder teeth.
4. At each end of the cylinder or rotor, measure the distance from a rasp bar to the concave. Measurements should be the same at both ends (Fig. 25).
5. Check for rasp bar wear. After long use, the center of the rasp bars may become worn so that equal clearance is not maintained along the concave. If rasp bars become worn to this extent, replace them.
6. Make sure correct spacing proportions between the cylinder and concave are maintained. Usually the front spacing is twice the rear spacing. Refer to the operator's manual for specific instructions.
7. Rotary combines may have uniform spacing between all concave bars and rotor rasp bars, or be set with wider space where crop enters the concave, depending on crop and threshing conditions. Refer to the operator's manual for specific instructions.
8. Move concave adjustment throughout full adjustment range daily to ensure free movement of linkage when adjustment is necessary.
9. Inspect the stone trap and empty it if necessary.
10. Check the speed of the primary countershaft to see it is adjusted to maintain the proper cylinder or rotor speed. Use a mechanical revolution counter (see Fig. 29).

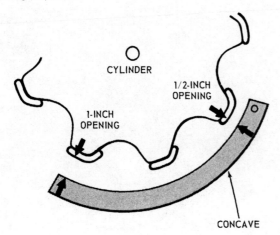

Fig. 26—Typical Front and Rear Cylinder-Concave Spacing Proportions

ADJUSTMENT

Typical adjustments of the grain handling units are described below.

Concave Leveling

If the concave is not level in relation to the cylinder or rotor, it must be adjusted. A leveling bolt is provided on each side of the concave. Adjust it as needed. The concave must be kept in correct relation to the cylinder or rotor (Fig. 26).

Rasp-Bar Cylinder

If Cylinder rasp bars are bent, remove and straighten them. On most combines, this can be done without removing the cylinder or rotor. To gain access, open the cylinder or rotor housing.

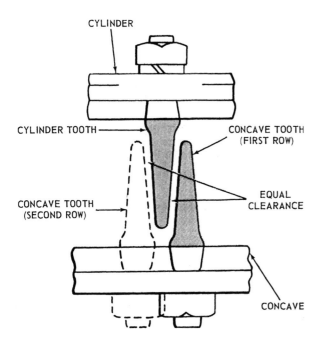

Fig. 28—Maintain Equal Clearance Between Cylinder And Concave Teeth

If rasp bars must be replaced, the mating bar which is located 180 degrees on the opposite side of the cylinder or rotor must be replaced to maintain proper cylinder or rotor balance (Fig. 27).

Spike-Tooth Cylinder

With the concave in high position and the teeth straight and properly adjusted, there should be equal clearance on either side of the cylinder teeth as they pass through the concave teeth (Fig. 28). On each side of the cylinder housing, most combines have adjusting bolts that move the concave from side to side as shown. Adjust these bolts so the teeth have equal clearance on either side. Turn the cylinder at least one full revolution to make sure all teeth have proper clearance. It may be necessary to straighten or replace individual teeth if proper clearance cannot be obtained.

Primary Countershaft Speed

The primary countershaft speed determines the basic cylinder speed. All drive speeds, except for ground speed, depend upon the proper primary countershaft speed. Therefore, it is vital that the drive for this shaft be kept in proper adjustment.

To check the countershaft speed, use a mechanical revolution counter — never guess (Fig. 29). If the speed is not correct, check the drive belts for proper tension, and adjust if necessary. Then check the countershaft speed again.

Always check countershaft speed with the engine at full throttle—no load.

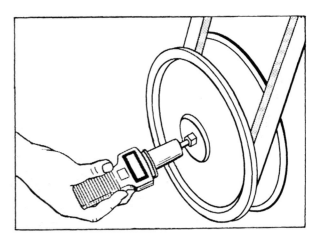

Fig. 29—Use a Mechanical Revolution Counter to Check Primary Countershaft Speed

If the speed is still incorrect, the engine speed needs adjustment. This is done by adjusting the governor on gasoline engines and injection pump linkage on diesel engines.

Refer to the operator's manual for specific instructions and speeds.

SEPARATING UNIT MAINTENANCE

Typical maintenance of the separating unit is described below.

LUBRICATION

Most separating units are designed with sealed bearings that do not require lubrication. For those parts that do require periodic lubrication, check the operator's manual.

INSPECTION

1. Check all belts and chains for wear and proper adjustment. Refer to belt and chain maintenance procedures later in this chapter.

2. Inspect the slip clutches and the separator drive unit to see if they are adjusted properly.

3. Check the straw walkers on conventional combines — rotary combines have no straw walkers. Make sure they are clean and are not bent or broken (Fig. 30).

4. Check the speed of straw walkers (if combine is so equipped and operator's manual recommends it).

5. Open inspection doors and clean out field trash.

6. Make sure grain conveyor augers are in good condition.

7. Check grain conveyor bevel gears for proper alignment.

8. Check the grain conveyor chain adjustment.

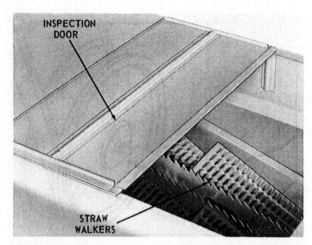

Fig. 30—Checking Straw Walkers through Inspection Door

ADJUSTMENT

Make the following adjustments, if necessary.

Grain Conveyors

The slip clutch spring should be set with just enough tension to work without slipping. Never increase the spring tension so much that the clutch will not slip.

Some straw walkers are equipped with shims for bearing adjustment. Avoid tightening the adjusting bolts too far. This can result in binding the bearing and damaging it.

To adjust the clutch, remove the cotter pin holding the adjustment nut in position and tighten or loosen the nut as needed. Set the clutch loosely at first and then tighten it gradually until the proper working tension is obtained.

Grain Augers

To adjust bevel gears, loosen the auger unit and move it up or down until the proper alignment is achieved.

To adjust the conveyor chain, lengthen or shorten the adjusting bolts as needed. Remember to adjust both sides equally.

CLEANING UNIT MAINTENANCE

LUBRICATION

Most cleaning units are designed with sealed bearings that do not require lubrication. For those parts that do require periodic lubrication, check the operator's manual.

INSPECTION

1. Check all belts and chains for proper tension and alignment. Refer to belt and chain maintenance procedures later in this chapter.
2. Inspect the fan blades periodically to see they are clean and straight.
3. Inspect the fan screen and clean it if necessary.
4. Check the fan speed with a mechanical revolution counter.
5. Check the chaffer for debris. Remove and clean it if necessary. Also check for bent or broken vanes.
6. Examine rubber bushings on the cleaning shoe for excessive wear or cracking. Remove and replace those in bad condition.
7. Inspect the sieve to see that screens are not plugged or broken.
8. Examine all bearings for excessive play.

ADJUSTMENT

When replacing worn or cracked rubber bushings, do not tighten fastening screws too tight. This may damage the new bushings. When installing a new bushing, it may be helpful to wet the shaft with water to act as a lubricant. Do not use oil, as it may damage the bushings.

GRAIN HANDLING UNIT MAINTENANCE

The grain handling units usually require the following maintenance.

LUBRICATION

Most grain handling units are designed with sealed bearings that do not require lubrication. For those parts that do require periodic lubrication, check the operator's manual.

Fig. 31—Inspect Elevator Doors and Remove Trash

Fig. 32—Inspect the Grain Tank

INSPECTION

1. Check all belts and chains for proper tension and alignment. Refer to belt and chain maintenance procedures later in this chapter.

2. Check elevator housings for breaks or weak points.

3. Open inspection doors on elevators. Check for trash and dirt and also inspect for broken or missing paddles (Fig. 31).

Inspect the grain tank (Fig. 32). Make sure it is kept dry and clean.

ADJUSTMENT

Typical adjustments of the threshing unit are described below.

Elevator Chains

Elevator chains are usually equipped with adjustment bolts on each side. These are normally located at the top or the bottom of the elevator. Adjust the bolts to obtain proper tension. Always make sure bolts on both sides are adjusted equally.

Slip Clutches

The slip clutch spring should be set with just enough tension to work without slipping. Never increase the spring tension so much that the clutch will not slip.

To adjust the clutch, remove the cotter pin holding the adjustment nut in position and tighten or loosen the nut as needed. Set the clutch loosely at first and then tighten it gradually until the proper working tension is obtained.

Unloading Auger

To remove the unloading auger, place the auger in the transport position (Fig. 33), and remove the cap screws or bolts holding it in position on the outer end of the auger. Then pull it from the tube. When replacing the auger in the tube, make certain it will properly align with the drive mechanism on the unloading auger.

Grain Tank

To clean or drain the grain tank, a door is usually provided at the bottom and to the outside of the bin (Fig. 34). Remove this door when cleaning or draining is necessary. When replacing the door, make sure it is fastened securely in the proper position to prevent grain leakage.

WHEEL AND TRACK MAINTENANCE

The following paragraphs describe maintenance of typical wheels and tracks.

Fig. 33—Place Unloading Auger in Transport Position Before Removing Auger From Tube

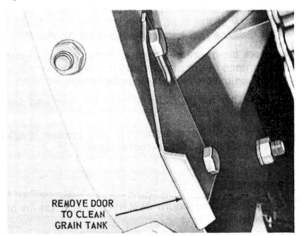

Fig. 34—Grain Tank Drain Door

LUBRICATION

Many *steering mechanisms* are equipped with sealed bearings at some points. However, almost all require some periodic lubrication. Refer to the operator's manual to make sure all points that need lubrication are serviced.

Most crawler tracks on rice combines are equipped with "button-head" grease fittings, which require a special gun for lubrication (Fig. 35). When greasing this type of bearing, refer to the operator's manual to

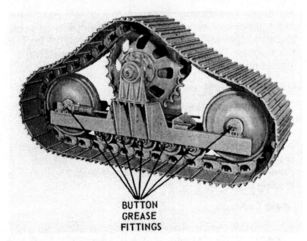

Fig. 35—Crawler Track for Rice Combine

determine the correct type of lubricant. Be especially careful not to over-lubricate these bearings. Seal damage can result and cause bearing failure.

INSPECTION

1. Check the rear steering wheels for correct toe-in which is required for easy steering and stable operation at higher speeds. This can be done by marking the center of the tread of each tire at the rear of the tire, (A, Fig. 36) in line with the axle. Measure the distance between these marks.

2. Move the combine forward until the marks on the tires are at the front, in line with the axle. Measure the distance between the marks again (B, Fig. 36).

   Check the operator's manual to determine what the difference in the measurement should be. If necessary, adjust the toe-in as detailed below.

3. Check the air pressure in both the rear and front tires.

4. If the combine is equipped with crawler tracks, check the tracks for correct tension.

ADJUSTMENT

The following procedure describes typical toe-in adjustments.

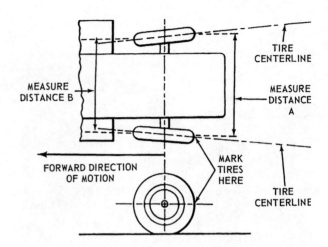

Fig. 36—Checking Toe-In of Wheels Which Steer Combine

Wheel Toe-in

The steering wheel toe-in can be adjusted by lengthening or shortening the tie rods. Loosen the jam nut on the tie rod and lengthen or shorten the tie rod as needed. Then tighten the jam nut securely.

 CAUTION: To adjust the rear wheel tread, jack up the combine and block it securely. Never rely on jacks alone for support.

Next, remove the pins in the telescoping axle and the tie rods and adjust to the width desired. Replace the pins in the proper holes.

Wheel Spacing

To widen front wheel tread on most combines, it is necessary to reverse the wheels. Always make sure the combine is blocked securely when reversing the wheels. Make sure proper clearance is maintained when wheels are reversed.

Some combines are equipped with wheel spacers. The width of the spacers varies but usually does not exceed four inches. When using spacers to widen or narrow the front tread, make sure the combine is blocked securely. It may be necessary to install dowel pins in these spacers to prevent shearing. Refer to the operator's manual for specific procedures.

Crawler Tracks

On combines equipped with crawler tracks, maintain proper track tension to prevent premature wear or machine failure. Most track mechanisms are equipped with large threaded bolts that adjust the tension. Tighten or loosen these bolts as needed. Some units are also equipped with adjustable idlers that can be moved in or out to obtain the proper tension. Consult the operator's manual for proper adjustment.

CARE OF RUBBER TIRES

Always keep tires off oil-soaked floors and away from fuel or oil. Oil softens and deteriorates rubber, shortening the life of tires. Bright sunlight causes the

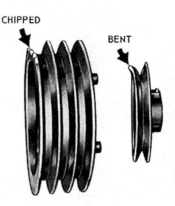

Fig. 38—Always Replace Belt Sheaves That Have Bent or Chipped Sidewalls

When checking the grease level in any gear housing, be sure to wipe away all dirt and grime so that none falls into the housing when the check or filler plug is removed. Always tighten filler plugs securely.

BELT AND CHAIN MAINTENANCE

Because many belts and chains are used on today's combines, proper maintenance is very important to keep them operating through the season.

Belts

The V-belts on a combine transmit power by friction and a wedging action against the sheaves. Belts are subject to increased wear through periodic heavy loads and should be checked often to be certain belt wear is normal. All belts and sheaves wear with use. Normal wear can be recognized as even wear—both on the belt and sides of the sheave.

A slight raveling or peeling of the belt at the lap does not indicate premature failure. Cut off the raveling if the covering peels at the lap.

Examine the sheaves for bent or chipped sidewalls (Fig. 38). Check for excessive sidewall wear. Damaged sheaves cause rapid belt wear. A bent sheave reduces the gripping power of the belt. Replace sheaves having any of the above defects.

Sheaves must be properly aligned to assure proper operation of all belts (Fig. 39). To check the alignment of sheaves use a straightedge. Place a straightedge or wire so that it runs from one sheave to the next. Then sight the alignment to see whether it is straight (Fig. 40). If it is not, adjust or replace the sheaves.

Make sure that dirt has not lodged and packed in sheave V-grooves. Excessive vibration may be caused by dirt collecting inside the sheaves. Loosen the dirt so it will fall out (Fig. 41).

Vibration can also be caused by lumpy V-belts. Check the belts for swelling and lumps.

Fig. 37—Always Wipe the Grease Fitting Clean Before and After Greasing

surfaces of tires to crack and harden. During storage, keep the combine out of the sun if possible.

If the combine will sit for over 60 days without being moved, jack up the axles to take the load off the tires.

Always refer to the operator's manual to determine the correct tire inflation pressures for the combine.

LUBRICATING WITH GREASE

Always wipe the grease fitting clean before placing the nozzle of the grease gun on the fitting (Fig. 37). This keeps foreign material from entering the lubrication point and damaging the bearing. After removing the grease gun, always wipe the fitting clean. Any grease left on the fitting will collect dirt and chaff.

If a fitting is missing, replace it immediately. Failure to do so will allow foreign matter to enter the bearing and could cause serious damage.

Never over-lubricate. Many bearings are equipped with seals that keep out dirt and abrasive material. Pumping too much grease into the fitting could cause this seal to rupture. Always use the recommended number of strokes or pump grease until increased pressure is felt.

Keep all lubricants in containers that protect them from dirt and moisture. Use care in filling the grease gun so that dirt does not enter the gun.

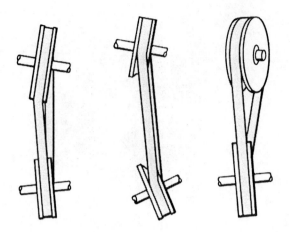

Fig. 39—Belt Sheave Misalignment

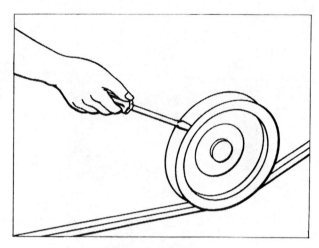

Fig. 41—Loosen Packed Dirt or Chaff in Sheaves

Always keep the belt clean. Oil or grease accumulating on the belts will cause them to deteriorate and slip.

Do not use belt dressings. Dressings often give only temporary gripping action while softening the belt and causing eventual deterioration, which shortens belt life.

Various types of tension devices are used for belt drives. For those with idler sheaves, the tension devices are found on belt drives. For belts equipped with idler sheaves, the tension may be adjusted by moving the idler sheave in or out (Fig. 42). For belt drives equipped with spring and bolt adjusters, increase or decrease the spring tension by adjusting the bolt (Fig. 43). Proper adjustment tension varies from drive to drive. Always check the operator's manual to determine the correct tension.

When it becomes necessary to replace a belt, first loosen the idler or tension adjustment bolts and then remove the belt. Never pry a belt on or off, as this can stretch and damage the belts and bend the sheaves (Fig. 44).

Chains

Chain drives normally have chain tighteners. Adjust chain tension so that chains are just tight enough to run without climbing or jumping sprockets and tight enough so that the chain doesn't whip or slap when running.

Be certain that all drive sprockets are properly aligned. Drive or driven sprockets may be moved in or out on shafts for proper alignment. Tightener sprockets may be aligned by increasing or decreasing the number of flat washers behind the sprocket hub.

Inspect the sprockets frequently to make sure teeth are not worn enough to cause damage to the chain.

Exposed chain drives should be cleaned regularly. Remove and clean them by soaking or dipping in diesel fuel. Dry and then oil the chain thoroughly after installing.

Fig. 40—Use a Straightedge to Determine Belt Sheave Alignment

Fig. 42—Idler Sheave Belt Tension Adjuster

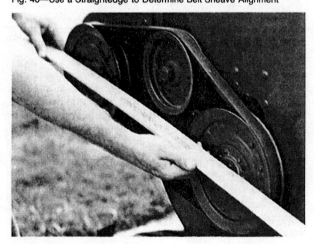

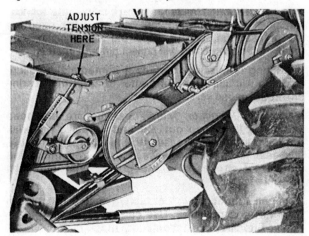

Fig. 43—Spring-Type Belt Tension Adjuster

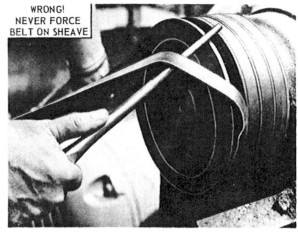

Fig. 44—Never Pry a Belt Onto or Off a Sheave

Before storing the combine, clean the chains and lubricate them with a heavy oil or grease. When removing the combine from storage, clean the chains again and lubricate them with motor oil.

PUTTING THE COMBINE INTO STORAGE

Proper storage of the combine will increase the life of the machine and save money on repairs (Fig. 45). If the combine cannot be stored inside, cover it with a tarpaulin.

Fig. 45—Storing The Combine

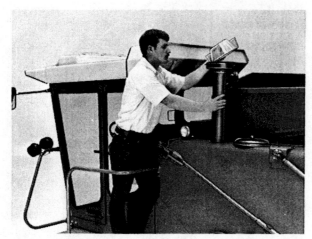

Fig. 46—Cleaning The Air Intake Precleaner

ENGINE

1. Wash the outside of the engine thoroughly. Use diesel fuel and a stiff brush.

2. Clean the inside of the air cleaner, remove any dirt from the filter and replace the filter in the air cleaner (Fig. 46).

3. Drain the crankcase while the engine is warm. Warm and agitated oil helps ensure that suspended dirt will drain with the oil rather than remain in the engine. Replace the oil filter and fill the crankcase with a new oil of proper viscosity and quality.

4. Drain, flush and refill the cooling system as follows:

- If freezing weather is anticipated, fill the system with antifreeze protection to a lower temperature than expected (Fig. 47). Half antifreeze and half water is recommended for most engines.

- If warm weather is anticipated, fill the system with clean water and a rust inhibitor, or use permanent-type antifreeze all year. Rust inhibitors are already mixed with the antifreeze.

Fig. 47—Checking the Radiator For Proper Antifreeze Solution

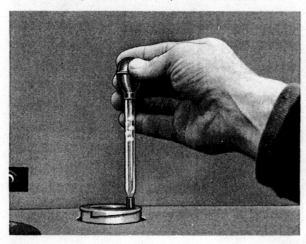

Fig. 48—Clean the Combine Thoroughly, Inside and Out

5. Drain the fuel tank. Also drain the fuel sediment bowl.

6. Drain the carburetor (gasoline engines only) by operating the engine at a fast idle until it stops. Add rust inhibitor to the fuel tank.

7. Add rust inhibitor to the crankcase.

8. Add rust inhibitor to the intake system. For a gasoline engine, pour rust inhibitor in each spark plug opening. For a diesel engine, remove the intake pipe running to the manifold and pour rust inhibitor in each cylinder. Replace the hose and turn the engine over slowly for two revolutions. Do not allow the engine to start.

9. If the engine is equipped with a turbocharger, disconnect the unit from the manifold intake pipe. Pour rust inhibitor into the manifold. Align the air intake pipe with the manifold and the turbocharger and tighten the clamps. Then turn the engine slowly for two revolutions.

10. Remove the batteries and store them in a cool dry place, not subject to freezing. Check the battery in storage every 30 days and recharge it if necessary.

COMBINE UNIT

1. Shelter the combine in a dry place.

2. Clean the combine thoroughly inside and out (Fig. 48). Chaff and dirt will draw moisture and rust the machine.

IMPORTANT: DO NOT use high pressure washer spray directly on bearings or any area that can be damaged by water. High pressure water can get past most seals and cause damage. If these areas get wet, dry them immediately. Then lubricate them and run the combine.

3. If possible, remove all belts, clean them and store them in a cool, dry place away from sunlight. If the belts are left on the machine, release the tension and place paper between them and any metal parts (Fig. 49).

4. Clean out the augers and elevators. Clean out the grain tank and open the drain hole cover.

5. Clean the chaffer and sieve.

6. Grease the feeder house conveyor bottom so it will not rust.

7. Drain the transmission, final drives and the hydraulic system. Then refill with the proper fluids.

8. Add rust inhibitor to the transmission, hydraulic system and final drives.

9. Lubricate the combine completely. Grease the threads on bolts used for adjustments and apply a coating of grease to all slip clutch jaws.

10. Paint all parts from which the paint has been worn.

11. Support the platform or corn head with blocks to level it.

12. Block up the combine to take the load off the tires.

13. Release spring tension on slip clutches.

14. Do not store hydraulic cylinders in the extended position. Heat may cause the oil to expand and burst the cylinders.

15. Make a list of the repairs that will be needed before the next season and order the parts immediately. These parts can be replaced in your spare time so the combine will be ready for the next harvesting season.

REMOVING THE COMBINE FROM STORAGE

Taking the combine from storage to ready it for harvesting involves several steps. Besides the steps mentioned here, make a thorough inspection of the combine to make sure all parts are in good operating condition.

Check the engine, transmission, final drives and the hydraulic system for leaks and add fluids as needed.

ENGINE

1. Remove the radiator cap and check the coolant level. If water has been used for storage purposes, drain the system and replace the water with the proper coolant. Check for loose hose connections and other leaks.

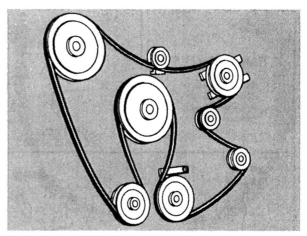

Fig. 49—Loosen All Belts If Stored On The Machine In Off Seasons

2. Clean the inside of the air cleaner and install a new element if the combine is equipped with a dry filtering system.

3. Remove the engine rocker arm covers and check to see that valves open and close freely by turning the engine over slowly. If the valves seem sticky or if excessive rust or moisture has formed, squirt upper cylinder lubricant on the valve stems to free them.

4. Install batteries and check the electrolyte levels. Recharge the batteries if necessary.

5. Clean and adjust the spark plugs. Replace worn or oil soaked wiring and check for cracked or peeling insulation.

6. Clean all the fuel lines and fuel filters. Blow out the carburetor jets with air.

7. For diesel engines, bleed the entire fuel system before starting the engine.

COMBINE UNIT

1. Clean the combine thoroughly, inside and out. Wash off dust and clean out any nests made by birds, insects or rodents.

2. Remove paper from between belts and metal parts and adjust the belts to the proper tension.

3. Adjust chains to the proper tension.

4. Clean all the slip clutches and adjust to the proper setting.

5. Replace the drain door at the bottom of the grain tank.

6. Lubricate the combine completely.

7. Check to be sure all nuts and bolts are tight.

8. Start the combine and run it at half speed for several minutes. Check all bearings for over-heating or excessive looseness. Be sure all slip clutches are operating freely. Check for leaks and loose connections.

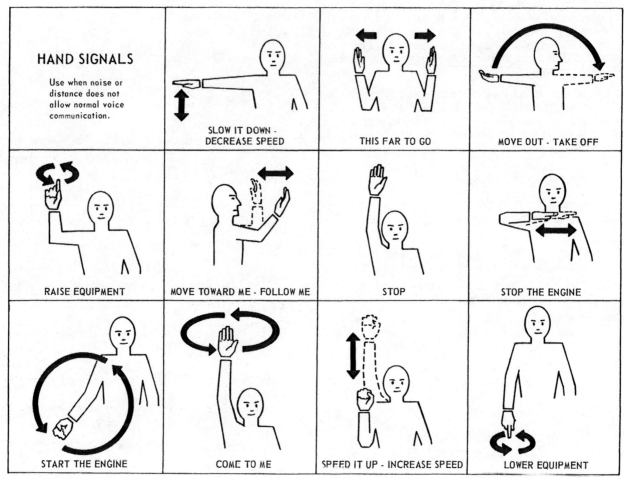

Fig. 50—Accepted Hand Signals

SAFETY

Good safety habits are a must for anyone who operates or services one of today's combines. Every year engineers and technicians develop safety devices that make combines safer to operate, but the responsibility for safety still remains with the operator. The operator must be aware of hazards and remain alert to situations that are potentially dangerous.

HAND SIGNALS

Spoken signals are very difficult to hear over the sounds of a modern combine. Hand signals from a person on the ground can be helpful to the operator when maneuvering a combine in tight spaces. Several safety institutions have endorsed a set of universal signals, as shown in Fig. 50. Study these symbols and use them to prevent accidents.

SAFETY BEFORE STARTING

Before attempting to operate a combine, study the operator's manual. It has information on general safety rules, plus specific safety recommendations for the particular machine. The more you know about the combine, the better prepared you will be to safely operate it.

The exhaust fumes from a gasoline or diesel engine are very poisonous. If the combine is run inside, be sure to open the doors to provide good ventilation.

Always clean the combine before starting. Trash around the exhaust system can cause fires. Oil, grease or mud on ladders or the platform can cause serious falls. If the combine is equipped with a cab, clean the glass to provide safe visibility.

Check the tire pressure each day (Fig. 51). Under-inflation can cause buckling of the sidewall, which can cause dangerous tire failure. Over-inflated tires have a great deal of "bounce" and cause upsets more readily than tires with correct pressure.

Check the brakes once a week. With hydraulic brakes,

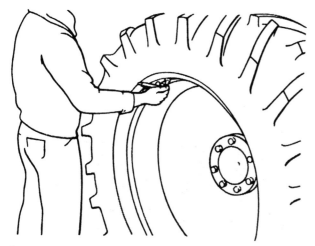

Fig. 51—Checking the Tire Pressure

make sure that the master cylinder is full of fluid and that no air is present in the lines. Adjust the pedal free travel, if necessary, so that the brakes are engaged with the pedals an equal distance from the floor of the platform. Check the operator's manual for specific instructions.

Check the threshing cylinder rocking bar to see it is clear of the cylinder.

Make sure that all shields and covers are in place and fastened securely.

Remove or stow all service equipment.

Always use the handrails and ladders provided on the combine for safe mounting and dismounting (Fig. 52).

STARTING THE COMBINE

1. Before mounting the combine, make sure that everyone is clear of the machine. Do not allow anyone to ride with you, unless combine is equipped with a passenger seat.

2. Before starting the combine:

- *Disengage header drive*
- *Disengage separator drive*
- *Place gearshift in neutral*
- *Depress clutch pedal*

3. Be careful when using diesel starting fluid. It is extremely flammable.

4 If it is necessary to use jumper cables to start the combine, be careful to avoid sparks around the battery. Hydrogen gas escaping from the battery can explode. Follow the operator's manual instructions for using jumper cables.

TRANSPORTING THE COMBINE

Always keep your mind on the dangers of driving the combine on public roads. Besides maintaining control of the machine, you must watch for obstacles on the road, pedestrians and traffic.

High speed is the leading cause of accidents. Never drive faster than road conditions allow for safe operation. Anticipate dangers and slow down to avoid accidents.

Make sure you are familiar with local traffic laws. Check the safety flashers and SMV emblems to be sure they are clean and visible.

Always lock the brake pedals together (Fig. 53). If the combine is not equipped with a locking mechanism, be sure to depress both pedals at the same time *evenly*. Applying only one brake, or applying one harder than the other can cause the combine to swerve and perhaps tip over.

Be careful when applying brakes when a header is attached to the combine. The added weight up front

Fig. 52—Use Handrails and Ladders For Safe Mounting And Dismounting

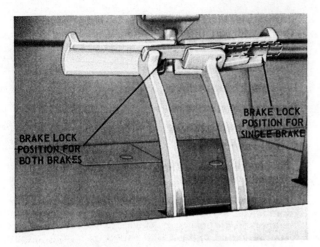

Fig. 53—Lock Brake Pedals Together When Transporting Combine

Fig. 54—Unloading Auger in Transport Position

can cause the combine to tip forward if the brakes are applied abruptly. Always drive slow enough to allow controlled application of brakes at all times.

Always check headlights and safety flashers to make sure they are properly adjusted and in working order.

Put the unloading auger in the transport position (Fig. 54). Be certain it is not blocking a safety flasher or SMV emblem.

On self-propelled combines, *never* use the header safety support when transporting the machine. Raise the header enough for safe ground clearance, but not high enough to reduce visibility.

On pull-type combines, always use header support when transporting. Towing at transport speeds can be hazardous because of side forces on the tractor when stopping too quickly. Side forces from slowing a combine too quickly may cause a tractor to skid, especially on loose gravel. Slowing down while turning can cause jackknifing. Slow down before the corner so the towed combine doesn't get out of control.

Fig. 55—When Transporting Over Long Distances, It is Safer to Haul the Combine

Fig 56—Transmission In "Tow" Position

Watch for low power or telephone lines, bridges, buildings and any other obstacles, to make sure you can pass under them safely. Always keep as far to the right of the roadway as possible. Keep a careful watch to see that you have safe clearance on both sides.

Always sit down when traveling at high speeds or going over rough terrain.

Be careful when making turns. Make sure that the rear of the combine will clear obstacles when it swings around. Avoid sharp turns. Turning too sharply at high speeds can cause the machine to turn over.

Because the wheels for steering are in the back, self-propelled machines often fishtail when turned too quickly at transport speeds. Steering to the right will whip the rear to the left, and vice versa. Steering suddenly to the right when meeting oncoming traffic causes the back of the combine to swing out into the path of oncoming traffic.

Slowing or braking too rapidly could cause loss of some steering control (weight on rear wheels). This is most noticeable when driving with a corn head or some other heavy header raised high. In this case, most of the weight will be on the drive wheels. Install rear wheel weights. Keep header as low as possible. Use the variable speed drive or engine throttle to slow the machine. Reduce speed before you need to apply brakes and always lock brake pedals together.

Never depress the clutch pedal or take the combine out of gear to coast down hill. When the combine is moving it is impossible to shift the transmission back in gear. Always maintain complete control of the combine. The same applies to tractors that are towing pull-type combines.

TOWING THE COMBINE SAFELY

If the combine must be transported over long distances, it is safer to haul it on a large truck or a special low trailer (Fig. 55)

Never tow the combine at speeds higher than 20 mph.

Always keep the transmission in neutral or in the "tow" position, if the combine is so equipped (Fig. 56).

Never tow a combine equipped with hydrostatic drive. Towing can cause damage to the drive unit. Instead, haul the combine.

OPERATING SAFETY

Never operate the combine if you are ill or sleepy.

Fig. 57—Be Very Careful When Operating on Hillsides

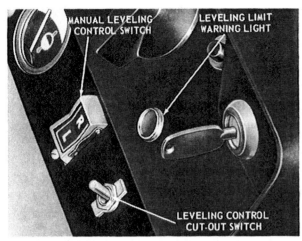

Fig. 58—Leveling Controls For Hillside Combine

Fig. 59—Always Disengage PTO Before Dismounting Tractor or Servicing Pull-Type Combines

Operating safety depends on alert, efficient handling of the combine.

Wear safety glasses at all times.

Wear clothing that fits snugly to avoid catching clothing in moving parts.

Never let anyone ride on the combine unless it is equipped with a passenger seat. A rider's clothing may become entangled in moving parts, or he may be thrown off the machine.

Before starting to harvest a field, check it carefully for ditches, fences or other obstacles. Be aware of weather conditions which present safety hazards.

Be especially careful when operating on hillsides (Fig. 57). Avoid sharp turns that could tip the combine over. Beware of ditches or obstacles — they are doubly dangerous on slopes.

If grain tank extensions are used, remember that the added weight may make the combine top heavy and more subject to upsets.

Never travel over 10 mph (16 km/h) with a full grain tank. The added weight makes the combine more difficult to maneuver and easier to upset.

Always sit down when traveling over rough terrain. A sharp jolt can throw you from the platform or away from the controls.

Hillside combines are equipped with automatic or manual leveling devices (Fig. 58). Hydraulic cylinders act to level these combines on steep slopes. These machines are equipped with a warning signal that indicates when the leveling system has reached its limit. Be especially careful after the device activates.

When using the steering brakes, always turn the steering wheel *before* applying the steering brakes. Failure to do so can cause the combine to swerve and turn dangerously.

FIELD REPAIR AND MAINTENANCE SAFETY

Always keep the machine clean. Field trash around the exhaust system can cause fires. Mud, grease or oil on the operator's platform or ladders can cause falls.

Before lubricating or adjusting the combine, disengage all drives and stop the engine. Never leave the operator's platform with the engine running.

Make sure that the header drive and separator drive are disengaged before attempting to clean the combine. Never try to unclog the machine with a stick or pole while the machine is running. The stalk rolls on a corn head can pull a 12 foot (3.6 cm) stick through in one second — shorter sticks or stalks even faster — before you can let go.

On pull-type combines, always disengage the PTO and turn off the tractor before attempting to unclog, adjust or lubricate the machine (Fig. 59).

Always stop the machine before opening inspection doors.

Keep all shields in place. After working on the combine, make sure the shields are fastened securely.

When operating in very dusty or noisy locations, wear goggles and ear plugs to insure safe visibility and prevent hearing loss. Never wear loose clothing that can become entangled in moving parts.

Stay clear of moving parts at all times.

Keep belts and chains properly adjusted and aligned.

Don't rely on the hydraulic system for support when working under the machine header. Always use the stops or supports provided on the machine. If no safety device is provided, block the header securely.

When adjusting the wheel spacing, make certain the machine is blocked. Never rely on jacks alone for support.

205

Fig. 60—Be Careful When Refueling Combine

Always support the reel arm securely when adjustments are being made.

Be careful when removing heavy parts. Make certain they are held firmly to avoid dropping them. Have someone help you with heavy jobs.

When operating in dry fields, install a spark arresting muffler to prevent fires.

Avoid sparks or open flames when working with the battery. Hydrogen gas escaping from the battery may explode.

When possible always refuel the combine outside the field. Let the engine cool before attempting to refuel and never smoke around fuels (Fig. 60).

Allow the system to cool and remove the radiator cap slowly, turning it until pressure escapes through the overflow pipe (Fig. 61). Make sure all pressure is relieved before removing the cap.

Stay clear of the exhaust system until it cools.

High-pressure fluid leaks in the hydraulic or diesel fuel system are very dangerous. The leaks can be invisible and still have enough pressure to penetrate the skin. When checking for leaks, use a piece of cardboard. If an injury does occur, seek medical aid immediately.

Always carry a first aid kit and fire extinguisher on the combine.

STOPPING THE COMBINE SAFELY

To make sure drive units do not cause injury when the machine is started again, do the following when stopping the combine.

1. Disengage header drive
2. Disengage separator drive
3. Place gearshift lever in neutral
4. Lower header

Fig. 61—Always Remove the Radiator Cap Slowly

5. Apply parking brake
6. Remove ignition key to prevent tampering or accidental starting.

REMEMBER: The hydrostatic drive unit is *not* an effective parking brake.

AVOID HIGH-PRESSURE FLUIDS

Escaping fluid under pressure can penetrate the skin causing serious injury. Avoid the hazard by relieving pressure before disconnecting hydraulic or other lines. Tighten all connections before applying pressure. Search for leaks with a piece of cardboard. Protect hands and body from high pressure fluids.

If an accident occurs, see a doctor immediately. Any fluid injected into the skin must be surgically removed within a few hours or gangrene may result. Doctors unfamiliar with this type of injury should reference a knowledgeable medical source.

CHAPTER QUIZ

1. (Fill in blank.) All operators should set up a _____ chart to see that the combine receives adequate care.
2. (Fill in blank.) Chains should be _____ when they are lubricated.
3. (Fill in blank.) Whenever a person works under a combine header, the _____ stop should be used.
4. True or false? "Some bearings do not need periodic greasing."
5. True or false? "When lubricating grease fittings, pump grease gun until grease appears at seals."
6. (Fill in blank.) Bent or chipped _____ will cause excessive sidewall wear on V-belts.
7. True or false? "A safe machine is more important than a safe operator."
8. Match the hand signal commands with the proper description below.

___1. Stop the engine   ___A. Hands held apart in front of face
___2. Start the engine   ___B. Arm and finger pointing down with circular motion
___3. This far to go   ___C. Hand motion back and forth across throat
___4. Lower equipment   ___D. Circular cranking motion of hand

Suggested Readings
Appendix
Index

SUGGESTED READINGS

TEXTS

Farm Power and Machinery Management: Hunt; Iowa State University Press; 1973.

Combines and Combining; Ohio State University, 1981.

Machines for Power Farming; Stone and Gulvin; John Wiley and Sons, Inc., New York, 1967.

Principles of Farm Machines; Bainer, Kepner and Barger; AVI Publishing Co.; Westport, Connecticut, 1972.

Fundamentals of Service: Engines; John Deere Service Training, Dept. F., John Deere Road, Moline, Illinois 61265.

Fundamentals of Service: Electrical Systems; John Deere Service Training, Dept. F., John Deere Road, Moline, Illinois 61265.

Fundamentals of Service: Hydraulics; John Deere Service Training, Dept. F., John Deere Road, Moline, Illinois 61265.

Fundamentals of Service: Power Trains; John Deere Service Training, Dept. F., John Deere Road, Moline, Illinois 61265.

Fundamentals of Service: Air Conditioning; John Deere Service Training, Dept. F., John Deere Road, Moline, Illinois 61265.

Fundamentals of Service: Fuels, Lubricants and Coolants; John Deere Service Training, Dept. F., John Deere Road, Moline, Illinois 61265.

Fundamentals of Service: Tires and Tracks; John Deere Service Training, Dept. F., John Deere Road, Moline, Illinois 61265.

Fundamentals of Service: Belts and Chains; John Deere Service Training, Dept. F., John Deere Road, Moline, Illinois 61265.

Fundamentals of Service: Bearings and Seals; John Deere Service Training, Dept. F., John Deere Road, Moline, Illinois 61265.

Fundamentals of Service: Shop Tools; John Deere Service Training, Dept. F., John Deere Road, Moline, Illinois 61265.

Fundamentals of Machine Operation: Preventive Maintenance; John Deere Service Training, Dept. F., John Deere Road, Moline, Illinois 61265.

Fundamentals of Machine Operation: Tractors; John Deere Service Training, Dept. F., John Deere Road, Moline, Illinois 61265.

Machinery Management: John Deere Service Training, Dept. F, John Deere Road, Moline, Illinois 61265.

Fundamentals of Machine Operation: Agricultural Machinery Safety; John Deere Service Training, Dept. F., John Deere Road, Moline, Illinois, 61265.

Conservation Farming; John Deere Service Training, Dept. F., John Deere Road, Moline, Illinois 61265.

VISUALS AND FILMS

Combine Harvesting Slide Set. 35 mm color. Matching set of 200 slides for illustrations in FMO Combine Harvesting text. John Deere Service Training, Dept. F., John Deere Road, Moline, Illinois 61265.

TEACHER'S GUIDE AND STUDENT WORKBOOK

FMO Combine Harvesting Teacher's Guide (FMO-15503T). Activities, exercises and transparency masters based on text. Student Workbook (FMO-15603W). Study questions. John Deere Service Training, Dept. F, John Deere Road, Moline, Illinois 61265.

SOFTWARE

FMO Combine Harvesting software and other agricultural software. Write for Software Catalog. John Deere Service Training, Dept. F, John Deere Road, Moline, Illinois 61265.

APPENDIX

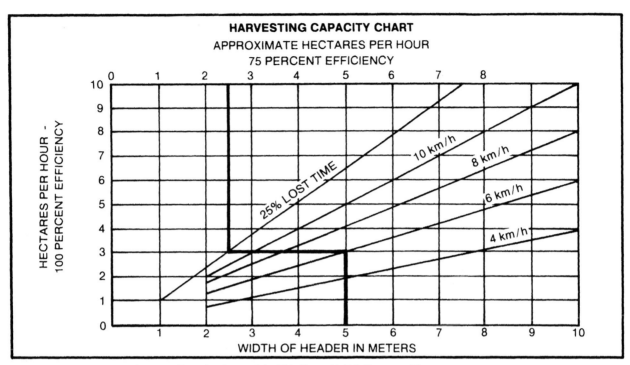

Combine Harvesting Hectares-Per-Hour Chart (See Page 132 for This Chart with U.S. Customary Measurements)

MACHINE-KERNEL LOSS CHART FOR CORN						
(Approximate Kernels Per Square Meter To Equal 50 Kilograms Per Hectare)						
Corn Head Size	Row Spacing (cm)	Separator Width (cm)				
		97	112	140	152	165
3-Row	90	44	38	—	—	—
	95	46	40	—	—	—
	100	49	42	—	—	—
4-Row	70	45	39	—	—	—
	75	49	42	—	—	—
	80	52	45	—	—	—
	90	58	51	40	37	—
	95	62	53	43	39	—
	100	65	56	45	41	—
5-Row	70	—	49	39	—	—
	75	—	53	42	—	—
	90	—	63	51	47	43
	95	—	67	53	49	45
	100	—	70	56	52	48
6-Row	70	—	59	47	44	40
	75	—	63	51	47	43
	90	—	76	61	56	52
	95	—	80	64	59	54
	100	—	84	68	62	57
8-Row	70	—	79	63	58	53
	75	—	84	68	62	57
	90	—	—	81	75	69
	95	—	—	86	79	73
	100	—	—	90	83	76
12-Row	70	—	—	94	87	80
	75	—	—	101	93	86

Machine Kernel Loss Chart For Corn (See Page 163 for This Chart with U.S. Customary Measurements)

MACHINE LOSS CHART FOR SMALL GRAIN											
(Approximate Number Of Kernels Per 1/10 Square Meter To Equal 50 Kilograms Per Hectare)											
Crop	Separator Width (cm)	Cutting Width (Meters)									
		3	4	4.3	4.6	4.9	5.5	6.1	6.7	7.3	9.1
Barley	97	54	71	77	82	87	98	109	—	—	—
	112	46	62	66	71	76	85	94	103	—	—
	140	—	49	53	57	61	68	75	83	90	112
	152	—	46	49	52	56	63	69	76	83	104
	165	—	—	—	48	51	58	64	70	77	95
Beans-Red Kidney	97	4	5	6	6	7	7	8	—	—	—
	112	3	5	5	5	6	7	7	8	—	—
	140	—	4	4	4	5	5	6	6	7	8
	152	—	3	4	4	4	5	5	6	6	8
	165	—	—	—	4	4	4	5	5	6	7
Beans-White Pea	97	6	9	9	10	11	12	13	—	—	—
	112	6	8	8	9	9	10	11	13	—	—
	140	—	6	6	7	7	8	9	10	11	14
	152	—	6	6	6	7	8	8	9	10	13
	165	—	—	—	6	6	7	8	9	9	12
Oats	97	43	58	62	66	71	79	88	—	—	—
	112	38	50	54	58	61	69	76	84	—	—
	140	—	40	43	46	49	55	61	67	73	91
	152	—	37	40	42	45	51	56	62	67	84
	165	—	—	—	39	42	47	52	57	62	77
Rice	97	56	74	80	85	91	102	113	—	—	—
	112	48	64	69	74	79	88	98	108	—	—
	140	—	51	55	59	63	71	78	86	94	117
	152	—	47	51	54	58	65	72	79	86	108
	165	—	—	—	50	53	60	67	73	80	99
Rye	97	97	130	140	149	159	179	198	—	—	—
	112	84	112	121	129	138	155	172	188	—	—
	140	—	90	97	104	110	124	137	151	164	205
	152	—	83	89	95	102	114	126	139	151	189
	165	—	—	—	88	94	105	116	128	139	174
Sorghum	97	51	68	74	79	84	94	104	—	—	—
	112	44	59	64	68	73	90	99	108	—	—
	140	—	47	51	55	58	65	72	79	87	108
	152	—	44	47	50	54	60	67	73	80	99
	165	—	—	—	46	49	55	61	67	73	91
Soybeans	97	9	12	13	14	15	16	18	—	—	—
	112	8	10	11	12	13	14	16	17	—	—
	140	—	8	9	10	10	11	13	14	15	19
	152	—	8	8	9	9	10	12	13	14	17
	165	—	—	—	8	9	10	11	12	13	16
Wheat	97	48	64	68	73	78	87	97	—	—	—
	112	41	55	59	63	67	76	84	92	—	—
	140	—	44	47	51	54	60	67	74	80	100
	152	—	41	44	47	50	56	62	68	74	92
	165	—	—	—	43	46	51	57	63	68	85

Machine Loss Chart For Small Grain (See Page 159 for This Chart with U.S. Customary Measurements)
Row Length Chart (See Page 162 for This Chart with U.S. Customary Measurements)

ROW LENGTH IN METERS PER 1/200 HECTARE							
Row Width- cm	One Row	Two Rows	Three Rows	Four Rows	Six Rows	Eight Rows	Twelve Rows
50	100	50	33.3	25	16.7	12.5	8.3
70	71.4	35.7	23.8	17.8	11.9	8.9	5.9
75	66.7	33.3	22.2	16.7	11.1	8.3	5.6
90	55.6	27.7	18.5	13.9	9.3	7	4.6
95	52.6	26.3	17.5	13.2	8.8	6.6	4.4
100	50	25	16.7	12.5	8.3	6.2	4.2

INDEX

A

Adjustments, Cleaning Units	153
Adjustments, Field	142
Adjustments, Header units	143
Adjustments, Importance of	121
Adjustments, Threshing units	151
Adjustments, Windrower	128
Air Conditioner controls	115
Angle-bar cylinder and concave	37
Auger, Adjustments, corn head	150
Auger, Adjustments, platform	145
Auger, Clean grain	55
Auger, Operation (platform)	31
Auger, Tailings	56
Auger, Unloading	56
Automatic header height control	30
Axles, drive, types	81

B

Beater and finger grate	43
Beater drive	76
Belt maintenance	195
Bin, grain	55
Brake, parking	102
Brake pedals	101

C

Cab, operator's	115
Capacities of combines, production	133
Chaffer	52
Chaffer adjustment	154
Chaffer controls	110
Charging circuit, engine electrical system	69
Chopper drive, straw	77
Clean grain elevator and augers	55
Clean grain elevator and augers, drives	77
Cleaning fan	50
Cleaning fan controls	110
Cleaning fan drive	77
Cleaning section mechanical problems	136
Cleaning shoe	51
Cleaning shoe drive	76
Cleaning shoe supply conveyor drives	76
Cleaning the crop	49
Cleaning unit	50
Cleaning unit controls	110
Cleaning unit maintenance	192
Cleaning units, operating	153
Clutch, power train	78
Combine field operation, controlling	106
Combine mechanical condition	134
Combine preliminary settings	137
Combine production capacities	133
Combine storage	197
Combining, when to	132
Compression stroke, engine	64
Concave adjustments	39
Concave and cylinder (rotor) function	38
Concave and cylinder (rotor) settings (effects of)	41
Concave, angle bar cylinder	37
Concave, rasp-bar	35
Concave spacing controls	109
Concave, spike-tooth	36
Concave-to-cylinder (or rotor) spacing	38
Conditions, lodged crop	166
Condition of combine, mechanical	134
Conditions, weather	165
Conditions, weedy field	164
Controlling combine field operation	106
Controlling combine movement	97
Controls and instruments, engine	96
Controls, identifying	94

C (continued)

Controls, operating	93
Cooling system, engine	69
Corn head adjustments	147
Corn head controls	106
Corn head drive	75
Corn head ear losses, how to determine	161
Corn head operation	31
Corn, how to determine losses	161
Crop conditions, lodged	166
Crop moisture content	167
Crop settings	138
Cutterbar, flexible floating	30
Cutterbar operation	29
Cutting and feeding the crop	25
Cutting platform adjustments	143
Cutting platform controls	107
Cutting platform drive	74
Cutting platform operation	25
Cylinder (rotor) and concave settings (effects of)	41
Cylinder, angle-bar	37
Cylinder, hydraulic (header lift)	89
Cylinder, hydraulic (reel lift)	88
Cylinder, hydraulic (selective ground speed)	89
Cylinder, rasp-bar	35
Cylinder (rotor) speed controls, threshing	109
Cylinder (rotor) speed, threshing	39
Cylinder, spike-tooth	36
Cylinder stripper	37
Cylinder (rotor) threshing adjustments	39
Cylinder (rotor) types, threshing	35

D

Determining grain losses	154
Determining threshing action	42
Developments leading to modern combine	3
Diesel engine starting	98
Differential, power train	80
Draper operation	31
Drive axles	81
Drive system (power rear wheels)	87
Drives, final (power train)	81
Drives, header	73
Drives, hydrostatic	82
Drives, propulsion	73
Drives, separator	75
Driving the combine	100

E

Ear losses, corn, how to determine	161
Electrical accessories	112
Electrical power for accessories	76
Electrical systems, engine	69
Electrical system for leveling (hillside combine)	89
Elevator, clean grain	55
Elevator drives	77
Elevator, trailings	56
Engine compression stroke	64
Engine controls and instruments	96
Engine cooling system	69
Engine electrical systems	69
Engine fuel systems	65
Engine, fuel types	62
Engine gearcase operation	72
Engine governing systems	71
Engine horsepower ranges	13
Engine intake and exhaust systems	66
Engine intake stroke	64
Engine exhaust stroke	64
Engine lubricating system	69
Engine maintenance	183
Engine operation	63
Engine, operating	97
Engine, power stroke	64
Engine power, transmitting	71

213

E (continued)

Engine starting tips 100
Engine systems 65
Exhaust and intake systems, engine 66
Exhaust stroke, engine 64

F

Fan, cleaning 50
Fan controls 110
Fan drive 77
Fan speed adjustments, cleaning 153
Feeder conveyor adjustments 151
Feeder conveyor operation 33
Feeding and cutting the crop 25
Field adjustments, combine 142
Field adjustments, windrower 128
Field and highway lighting 112
Field conditions, lodged crop 166
Field conditions, weedy 164
Field operations and adjustments 142
Field operation controls 105
Field problems, troubleshooting 166
Field problems with windrowers 130
Final drives, power train 81
Flexible floating cutterbar 30
Floating cutterbar (flexible) 30
Fluid level system, hillside combine 89
Four Wheel Drive System 87
Fuel systems, engine 65
Fuel types of engines 62

G

Gasoline engine starting 99
Gatherer chain adjustments 148
Gatherer points, adjustment 148
Gear shifting, transmission 103
Grain handling units, maintenance of 192
Grain losses, determining 154
Grain losses, sources 155
Grain, small, how to determine losses 157
Grain tank 56
Grain tank unloading auger 56
Grain tank unloading auger controls 110
Grain windrowers 123
Ground speed controls 103
Ground speed, how to figure 134
Governing systems, engine 71

H

Hand signals, safety 200
Handling the crop 55
Harvester-thresher 2
Harvesting, crop settings 138
Harvesting factors 122
Harvesting methods 123
Harvesting planning and preparation 123
Harvesting systems 21
Harvesting, when to 132
Head, row-crop adjustment 150, 187
Header controls 105
Header drives 73
Header field operation and adjustments 143
Header height control (automatic) 30
Header losses 156
Header losses, how to determine:
    Corn 162
    Small grain 157
Header maintenance 185
Header mechanical problems 135
Header sizes 14
Header, types of 25
Heater controls, operator's cab 117

H (continued)

Hillside combine 11
Hillside combine leveling system 89
Hillside leveling controls 111
Horns, truck signal 114
How to measure losses 157
Hydraulic cylinder (header lift) 89
Hydraulic cylinder (hillside leveling) 89
Hydraulic cylinder (reel lift) 88
Hydraulic cylinder (selective ground speed) 89
Hydraulic drive motors 89
Hydraulic steering pump 89
Hydraulic system for leveling (hillside combine) 90
Hydraulic systems 87
Hydraulic drive controls 103
Hydraulic drives 82

I

Identifying controls 94
Ignition circuit (Spark-Ignition Engines) 70
Intake and exhaust systems, engine 66
Intake stroke, engine 64

L

Ladder, operator's 103
Leakage losses 157
Leveling controls, hillside combine 111
Leveling system (hillside combine) 89
Leveling system operation (hillside combine) 90
Level-land combines 11
Lighting, field and highway 112
Lodged crops 166
Losses, corn, how to determine 161
Losses, grain, sources 155
Losses, small grain, how to determine 157
Lubricating system, engine 69
Lubricating with grease 195

M

Machine losses, how to determine 157
Maintenance, belts 195
Maintenance, chains 196
Maintenance, cleaning unit 192
Maintenance, general 182
Maintenance, grain handling units 192
Maintenance, header 185
Maintenance, safe 204
Maintenance, separating unit 191
Maintenance, threshing unit 189
Maintenance, tires 194
Maintenance, wheel and track 193
Maintenance, windrower 129
Mechanical condition of combine 134
Moisture content, crop 165
Monitoring unit, shoe and straw walker losses 114
Monitoring unit, slow shaft speed 114
Motors, hydraulic 89

O

Operator's cab 117
Operator's ladder 103
Operator's seat adjustment 100
Operation, field 142
Operation, importance of 121
Operation of engines 63
Operation of windrowers 127
Operating and adjusting cleaning units 153
Operating and adjusting threshing units 151
Operating controls 93

Operating controls, symbols	116
Operating safety	204
Operating the engine	97
Overthreshing	151

P

Parking brake	102
Pickup platform controls	108
Pickup platform drive	75
Pickup platform operation	31
Planning and preparation	123
Platform auger operation	31
Platform controls, cutting	107
Platform controls, pickup	108
Platform, cutting, adjustments	143
Platform drive, cutting	74
Platform drive, pickup	75
Platform operating, cutting	25
Platform operation, pickup	31
Power stroke, engine	64
Power systems	59
Power train	77
Power train maintenance	183
Power train clutch	78
Preharvest losses	155
Preharvest losses, how to determine	157
Preparation and planning	123
Pressurizer fan switch, cab	117
Problems, combine field	167
Problems, combine mechanical	135
Problems, windrower field	130
Production capacities of combines	133
Propulsion drives	73
Propulsion unit controls	102
Pull-type combines	14
Pump, hydraulic steering	89

R

Racks, straw	43
Rasp-bar cylinder and concave	35
Reel adjustments	144
Reel fore/aft operation	28
Reel lift hydraulic cylinders	88
Repair safety	204
Rotary combine	5, 6, 25
Rotary separation	48
Rotor and concave	37
Rotor (cylinder) and concave adjustment	39
Rotor (cylinder) and concave function	38
Rotor (cylinder) and concave settings (effects of)	41
Rotor speed, threshing	39
Rotor (cylinder) threshing adjustment	39
Rotor types	35
Row-crop head adjustment	150, 187
Row-crop head operation	33

S

Safety, Hand signals	200
Safety, Operating	203
Safety, Starting	201
Safety, Stopping	205
Safety, Transporting	201
Seat adjustment, operator's	100
Selective ground speed hydraulic cylinder	89
Self-propelled combine types and sizes	4
Settings, crop	138
Settings, combine harvesting	137
Separating action, importance of	47
Separating the crop	43
Separating units, maintenance of	191
Separator controls	108
Separator drives	75
Separator mechanical problems	135

S (continued)

Separator sizes	13
Separator speed, effects of	41
Shifting gears	103
Shoe, cleaning	51
Shoe and straw walker losses monitoring unit	114
Shoe drive	76
Shoe losses	157
Shoe losses, how to determine:	
Corn	163
Small grain	160
Sidehill combines	11
Sieve	53
Sieve adjustment	154
Sieve controls	110
Signal horns, truck	114
Sizes of combines	12
Snapping plates, adjustment	149
Speed, combine separator	137
Speed, ground, how to figure	134
Special combines	16
Spike-tooth cylinder and concave	36
Spreader drive, straw	77
Stalk rolls, adjustment	148
Starting circuit, engine electrical system	70
Starting safety	201
Starting the diesel engine	99
Starting the gasoline engine	99
Starting tips, engine	100
Steering controls	101
Steering pump, hydraulic	89
Stone protection	33
Stopping the engine	99
Storage, combine	197
Storage, windrower	129
Straw chopper drive	77
Straw racks	44
Straw spreader drive	77
Straw walker	44
Straw walker action	46
Straw walker, separator, and shoe losses, how to determine:	
Corn	163
Small grain	160
Straw walker and shoe losses, monitoring unit	114
Straw walker drive	76
Straw walker losses	156
Straw walker plugging warning device	114
Stripper, cylinder	37
Symbols for operating controls	116

T

Tailings, elevator and augers (operation)	56
Tailings elevator and auger drive	77
Tank, grain	56
Thresher	2
Threshing action	41
Threshing action, how to determine	152
Threshing adjustments	39
Threshing cylinder sizes	14
Threshing cylinder (and rotor) types	35
Threshing controls	109
Threshing cylinder (rotor) and concave function	38
Threshing cylinder (rotor) drive	76
Threshing losses	156
Threshing losses, how to determine:	
Corn	163
Small grain	159
Threshing problems, how to correct	152
Threshing the crop	34
Threshing systems	35
Threshing units, maintenance of	189
Threshing units, adjustment of	151
Tire care	194
Track maintenance	193
Transmission	80
Transmitting engine power	71

T (continued)

Transporting safety	201
Transporting the combine	105
Transporting the windrower	129
Trash knives, adjustment	149
Troubleshooting field problems	166
Truck signal horns	114
Types and sizes of combines	4

U

Underthreshing	152
Unloading auger controls (grain tank)	110
Unloading auger drive	77
Unloading auger, grain tank	56

W

Walker, straw	44
Walker, straw, losses	156
Warning device, straw walker plugging	114
Weather conditions	165
Weedy field conditions	165
Wheel and track maintenance	193
Windrow pickup adjustments	146
Windrower field adjustments	128
Windrower operation	127
Windrower field problems	130
Windrower, grain	123
Windrower maintenance	129
Windrower storage	129
Windrower, transporting	129
Windrower, types and sizes	125
Windrowing, how to	127
Windrowing method of combining	15

COMBINE HARVESTING

Why do you need this book? The answer is simple: combine harvesting is **not** simple anymore. You probably learned how to "run" a combine at quite a young age. But the machines themselves keep getting more and more complex, so even operating them is no longer simple.

With all of the on-board monitors, computers, and now even satellite guidance systems, operating a modern combine is downright complicated! And it's not just the many controls, although they make the cab of today's combines look like the cockpit of a jumbo jet airliner. It's mastering the many tasks of field operation and skillful adjustments to the machine's components.

Without this mastery and skill, you wind up with:

- grain losses on the ground,
- unthreshed kernels on the straw, cob or in the pod,
- straw chewed up excessively,
- grain lost from the straw walkers or shoe,
- excessive tailings in the tailings elevator,
- cracked grain on the grain tank,
- chaff or trash in the grain tank;

and your farm ends up with profit losses because of marketing penalties for low-quality grain due to harvesting damage or crop condition.

But proper operation is just part of the solution. If you do not know your combine's production capacity; when to combine; how to determine grain losses and overcome them; or how to troubleshoot field problems – then you need to learn. And this book will teach you. And do it in easy steps. And you'll also learn how to make necessary machine adjustments...and how to maintain each component in peak operating performance.

This book shows you – with many illustrations – how to operate, maintain, and improve the efficiency of your combine. The harvesting capacity and machine loss charts in the back of this book and the troubleshooting charts throughout it are worth the purchase price in themselves.

Deere & Company

FMO15104NC
ISBN 0-86691-236-3
Litho in U.S.A. (J-00)